THE
SCIENCE
OF
FORTNITE

THE
SCIENCE
OF
FORTNITE

THE REAL SCIENCE BEHIND THE WEAPONS, GADGETS, MECHANICS, AND MORE!

JAMES DALEY

Skyhorse Publishing

Skyhorse Publishing books may be purchased in bulk at special discounts for sales promotion, corporate gifts, fund-raising, or educational purposes. Special editions can also be created to specifications. For details, contact the Special Sales Department, Skyhorse Publishing, 307 West 36th Street, 11th Floor, New York, NY 10018 or info@skyhorsepublishing.com.

Skyhorse® and Skyhorse Publishing® are registered trademarks of Skyhorse Publishing, Inc.®, a Delaware corporation.

Visit our website at www.skyhorsepublishing.com.

10 9 8 7 6 5 4 3 2 1

Library of Congress Cataloging-in-Publication Data is available on file.

Cover design by Brian Peterson
Cover photograph by gettyimages

Print ISBN: 978-1-5107-4962-7
Ebook ISBN: 978-1-5107-4963-4

Printed in the United States of America

Contents

Preface

Hi, parent. I say this because most people who play Fortnite are kids, and most kids don't read prefaces (if they read at all). If you are a kid and you've actually decided to read this preface of your own volition, congratulations—you are an even bigger nerd than me, and I just wrote a book about the science of Fortnite!

Anyway, whether you are a parent or a kid, I just want to say hello, welcome you to the book, and introduce you to the overall themes and concepts you'll find in the following chapters.

The aim of this book is simple: to explore the various science-fiction elements of Fortnite and try to find where they meet, overlap, and depart from actual scientific realities. To this end, we will explore the environment of the island where the game takes place, looking at the storm, the volcano, physics, and some strange otherworldly phenomena that you find all over the island. We will also look at some of the weird and wonderful technologies that exist in the Fortnite universe, such as Driftboards and jetpacks and the iconic Battle Bus. We will go into great detail about many of the weapons used in Fortnite, making sure to explain not only the scientific concepts behind how some of these weapons work, but also to discuss how understanding the science behind these weapons can help a player do better in the game.

While it is my sincere hope that you find this book entertaining, and that you do pick up a few things that help you get a few Victory Royales every once in a while, the real purpose of this book is to teach you a little about the fascinating science underpinning the world of Fortnite.

Many of the concepts explained in this book could easily be the subjects of their own books (and most of them are). Thus, I hope you will forgive any oversimplification in these pages, as it is necessary to provide a general understanding of these concepts to the casual reader and Fortnite player, hopefully without straying too far from the wacky, fun world of Fortnite that brought you here in the first place.

Enjoy!

—James Daley

PART 1
The Physics of Fortnite

GRAVITY

You might not think about it while too much while you're wandering around the island, trying not to get killed and searching for some sweet loot, but gravity in Fortnite, just like in every other video game, is one of the most important scientific concepts to understand if you really want to master the game.

Just like in real life, most people take gravity for granted in video games. If you jump up, you fall back down. If you accidentally slip and fall off a cliff, you drop until you hit the ground beneath you, and if that cliff is high enough (and you don't have that weird special ability where you glow a little and don't take any fall damage), you will die.

But why? What's actually happening when gravity is pulling you down toward the ground, both in Fortnite and in the real world? How is gravity different in Fortnite than it is in the real world, and how do the laws of physics apply—or not apply—in the Fortnite universe?

Before we get into all of that, we need to have a little history lesson. Gravity is one of the most interesting concepts in the history of science, and it all started with Sir Isaac Newton, and the apple that fell on his head while he was sitting under a tree . . . okay, not really. The story goes that Sir Isaac Newton was sitting under a tree when an apple fell on his head, and this made him decide to figure out what gravity is and how it works. In truth, this probably never happened. Much like Benjamin Franklin and the story of how he discovered electricity by flying a kite in a lightning storm, the story of Newton's apple is a nice way to illustrate a complex scientific principle, but it is most likely just a story. What is not just a story is the fact that Newton did, in fact, completely change the way humanity understood the world and the universe when he came up with his Universal Theory of Gravitation.

Though the apple-falling-on-the-head story is apocryphal, apples did play a role in the development of Newton's thinking about gravity. Newton had been interested in how gravity functioned even before he went to university, and this interest greatly informed his studies. He was especially intrigued by the role gravity played in the movement of stars and planets, and much of his academic work was focused in this area. Even with all of his scientific education, it wasn't until later in his life, when he observed some apples falling from a tree (no, they never actually landed on his head), that he first realized that the same force that kept apples from falling sideways (or even upward) could be responsible for the moon revolving around Earth, Earth revolving around the sun, etc.

What was it about that apple falling from the tree that gave Sir Isaac Newton the idea that eventually led to his Law of Universal Gravitation? Unlike the story, it wasn't merely the fact that the apple came down that made him wonder about the forces acting on it; it was the fact that this particular apple made him realize that *any* apple that fell from *any* tree at *any* spot on the surface of the Earth would fall in a slightly different direction than all of the others.

Now you're probably thinking, wait a minute, that doesn't make any sense! Any apple that falls out of any tree is going to fall down!

Well, yes. All apples do travel more or less straight toward the ground when they fall from a tree, but not all trees point in the same direction, do they? No, of course not. In fact, every tree is pointing in a slightly different direction, depending on where it is on the surface of the Earth.

Think about it this way: Imagine you have two apple trees planted on exact opposite sides of the Earth. Let's say one tree is in Wellington, New Zealand, and the other is just outside of Alaejos, Spain. Now, imagine that an apple were to fall from each of these two trees at the exact same time. What direction would each of these apples be traveling? Would the apples be going in the same direction? After all, they're both falling down toward the Earth, right?

No, of course they would not be going in the same direction. In fact, these two apples would be falling *directly toward each other*. If these apples somehow had the magical ability to pass through solid material, they would eventually collide at the exact center of the Earth . . . just like any object that falls from any other object at any point on the entire planet will *always fall toward the exact center of the Earth itself.*

This is what Sir Isaac Newton realized in the seventeenth century: some invisible force was pulling everything toward the very center of the Earth. All by itself, this realization was not really so revolutionary. After all, humans had known that the Earth was round for some time, so it stands to reason that someone (or even many people) would have reached this same conclusion. What made Newton's epiphany so special was that he didn't stop there. He extended this logic to the rest of the known universe, questioning whether these same principles could be applied to the moon, the planets, the sun, and possibly even all of the stars in the sky. What if each of these cosmic bodies exerted the same force from their centers that the Earth did? If true, what would that mean for the motion of the planets through the sky, or for the revolution of the moon around the Earth? Could one universal, invisible force really be acting upon every single thing in the universe?

It didn't take Newton long to realize that yes, such a force acting on everything would provide a much better explanation for the motion of the observable universe than anything that anyone had come up with previously. Newton deduced that the moon must be orbiting around the Earth because the Earth was generating a gravitational force on the moon, just like the Earth must be orbiting the sun because the sun was generating a greater gravitational force than the Earth.

But we're getting a little ahead of ourselves here. How did Newton get from an apple falling from a tree to a planet orbiting a star? If the Earth was exerting the same force on the apple as the sun was on the Earth, wouldn't the Earth just fall into the sun?

To illustrate his ideas, Newton devised this simple thought experiment. Imagine you have a cannon at the top of a tower (sure, let's make it a Pirate Tower; why not?) and you shoot that cannon precisely parallel to the ground below. What will the cannonball do? You don't need to be a seventeenth-century physicist and polymath to deduce that the cannonball will travel more or less in a straight line until gravity pulls it down to Earth. Obviously, the distance the cannonball travels depends on a lot of different factors (the height of the tower, the weight of the ball, the amount of gunpowder used, etc.), but let's say that the tower is 100 meters tall and that the cannonball travels 1,000 meters before coming to a stop on the ground.

Now, what do you think would happen if you took that same cannon, with the same cannonball, and the same amount of gunpowder, but you made the tower 200 meters tall instead of 100 meters, and again shot the cannonball precisely parallel to the ground. Would it travel the same distance as the first one, or farther? It would travel farther, of course. This much is obvious. But why?

The reason the cannonball travels farther is that it takes gravity longer to slow it down and to bring it back to Earth. Furthermore, the higher you place the cannon before shooting it, the less the force of gravity is exerting itself as it moves through the air.

So now, Newton asked, what happens when we build our tower even higher? What does the cannonball do if it's shot from 5 kilometers in the air, or 100 kilometers? Would the cannonball just keep on falling toward the earth no matter how high you built your tower?

For a while, yes. But what Newton deduced from his understanding of gravity was that, eventually, if you built the tower high enough (higher than any real tower could ever be), the cannonball would make it all the way around the Earth until it came back to the exact spot where the cannon fired. Furthermore, if you built your tower even higher than that, your cannonball would just keep on traveling around and around and around the Earth without ever falling all the way down. This is because, past a certain height, the force of gravity is only strong enough to change the direction of the cannonball, not strong enough to pull it all the way back down to the ground. This is because, at such a distance from the Earth, the amount of force that gravity is exerting on the cannonball is less than the force of the cannonball's forward momentum.

You can extend this even further and imagine a point where you could make the cannon so far from the surface of the Earth that gravity cannot make the cannonball do more than slightly bend on its trajectory before hurtling onward into space, never stopping at all.

These ideas, and this way of looking at the universe, remained humanity's best way of understanding its place in the cosmos for nearly five hundred years, until a young physicist named Albert Einstein came along and changed everything.

It's important to note that even though Newton's theories did prove themselves mostly accurate both mathematically and through experimentation (meaning that one could use his Universal Theory of Gravitation to accurately predict the way that gravity would function in the natural world), Newton never really figured out *why* gravity did all the things that he observed. In a sense, we still don't know everything about how gravity works, but Einstein's theory of relativity did manage to explain a lot that Newton's Universal Theory of Gravitation couldn't.

That said, we don't need to throw everything that Newton discovered out the window (or out of a tree, as it were) because quite a lot of his discoveries and innovations still hold true in light of Einstein's discoveries. After all, both Einstein and Newton were essentially looking at the same universe (though obviously, Einstein had more data at his disposal), and both came up with mathematically provable theories to explain what they observed in the natural world.

For example, it was Newton who figured out that gravity was responsible for the Earth and all of the other planets orbiting the sun, and Einstein certainly did not contradict that discovery at all. The difference between Einstein and Newton comes when you start to look at *why* gravity is doing the things it does. There were also certain circumstances in which Newton's mathematical models did not accurately describe everything that scientists had observed in the natural world by the time Einstein came along.

When you're talking about Einstein and gravity, it can be very easy to wander into all the other aspects of his general and special theories of relativity, but I am going to try to stick to gravity as closely as possible. And the best place to start, when talking about Einstein's theory of relativity and how it relates to gravity, is with the Equivalence Principle.

In simple terms, the Equivalence Principle states that it is impossible to distinguish between gravity and acceleration. But what the heck does that mean?

Imagine you wake up in an elevator, having no idea how you got there. As with most elevators, you can't see outside, you can't hear anything outside of the electronic hum of the elevator itself, and there's no real way of telling what is going on outside of your little box.

Now let's say that after you wake up in this strange elevator, you stand up and find that there's a rubber ball in your pocket. Being a bit bored, you decide to bounce the ball up and down on the floor of the elevator to pass the time. As far as you can tell, bouncing the ball in the elevator works the same as it always has any time you have bounced a ball anywhere on Earth.

Now let's say that your identical twin is in an identical elevator (at least, it looks identical from the inside), and everything is exactly the same, except that when your twin wakes up, he or she is floating in the air, completely weightless, with no sense of gravity whatsoever. When your twin tries to bounce the ball, it just hovers there, completely weightless.

Still with me? Good, now here's the interesting part: what if I told you that one of the two you is precisely where you expect to be—in an elevator, in a regular elevator shaft, in some tall building on the Earth somewhere—while the other one is actually in an elevator way out in space, a hundred million miles away from Earth? Would you have any way of telling for certain which one of you was on Earth and which one was in space?

Obviously, this is a trick question. The apparent right answer would be that the person in the weightless, seemingly gravity-free elevator is in space, while the other person in the elevator that seems to have normal gravity is on Earth. And obviously, if you were to figure out the probability of each scenario, that would probably be correct. However, the question was: can you tell *for certain* which twin is where?

The answer to this question is no. Why? Simple: The gravity on Earth is measured at 9.807 m/s^2 (meters per second per second). Now let's say that in the elevator on Earth is in an extremely long elevator shaft end it is accelerating toward the ground at exactly 9 m/s^2. Meanwhile, in space, let's say the other elevator has a booster rocket on the bottom of its frame, and is accelerating up (or what the twin perceives as "up") also at exactly 9 m/s^2. The twin inside the elevator accelerating toward Earth in the elevator shaft will feel as if there is no gravity whatsoever because the acceleration is exactly canceling out the gravity of the Earth. Meanwhile, the twin in the space elevator will be able to walk on the ground of the elevator and bounce a ball without being able to tell where they are, because the acceleration exactly matches the gravity of Earth. This illustrates the Equivalence Principle: there is no functional difference between gravity and acceleration.

The realization that gravity was exactly equivalent to acceleration caused Einstein to make one of the greatest discoveries in the history of science. He figured out the reason for this equivalence between acceleration and gravity: space-time is actually curved, and it's the mass of every object in the universe that curves it.

Here is another illustration to help you understand what this means: imagine a large rubber sheet that looks sort of like a loose trampoline attached at the edges to a circular frame. In this illustration, this rubber sheet is space-time. The fact that it is a two-dimensional sheet causes certain problems in thinking about it as space-time, but don't worry about that for now. Just imagine that space-time—that mysterious stuff that the entire universe is made of—is a big, loosey-goosey rubber trampoline.

Now let's say that you have a little yellow marble that is very small and very light. In this thought experiment, our little yellow marble is an asteroid coming from a distant part of the galaxy. This asteroid travels across this particular region of space-time at a very high speed, which we will illustrate by rolling the marble across the space-time trampoline. Now, being that the tiny little marble is the only thing on the sheet and it is not very heavy, it will roll more or less straight across. For the purpose of this experiment, let's assume that you are not rolling this ray-of-light marble through the exact center of the trampoline sheet, but a little off to one side.

Now we're going to add some mass to our space-time trampoline. Let's say you have a bowling ball that you've painted to look like the sun, and you place this bowling ball right square in the exact middle of your trampoline. What's going to happen to the rubber sheet? Well, obviously, it's going to sink down beneath the weight of the bowling ball, right? Right.

So at this point, what will happen if you roll that little marble on the same path that you had rolled it before you put the bowling ball on the space-time trampoline? Will it go straight across like it did the first time? No, of course not. As the marble approaches the section of the space-time trampoline sheet that is bent under the weight of the bowling ball, its trajectory will curve in the direction of the bowling ball itself. If the marble is going fast enough, it will only curve a little before continuing on its way to the edge of the trampoline, though at a slightly different location on the edge than where it went off the first time.

However, if you were to push that marble a little more slowly, or maybe even move it a little closer to the bowling ball itself, the curve in the marble's trajectory would increase, making it turn even further off of its original course. Continuing with this example, if you slow it down even more, and move it even closer to the bowling ball, eventually your little asteroid marble will not make it off the edge of the sheet at all. Eventually, the marble will curve all the way around and begin to circle the bowling ball until it eventually falls into it.

This is how gravity works with Einstein's theory of relativity. It is the curve in space-time that causes the gravitational force. When that apple Newton noticed first fell from the tree, it was falling because the mass of the Earth was bending space-time.

Mass actually bends space-time, causing motion to curve along with it, and the greater the mass, the greater the distortion in space-time. This is why the sun (which is very, very big) is the center of our solar system, and why the planets (which are smaller) orbit around it, and why the moon orbits around the Earth.

But how does this relate to the Equivalency Principle?

Let's say you take a really, really, really big rubber sheet, like the size of a football field, and you put one bowling ball one ten-yard line and another identical bowling ball the other ten-yard line. Will they begin to move toward each other? Probably not, because they are too far away for their curves to affect each other. They will each be in their own little pocket, not really affecting the other. If you begin to move these toward each other, however, eventually the pockets they are making in the trampoline will connect, and they will start to become attracted toward their counterpart.

What is particularly interesting is *how much* they become attracted to each other as they move closer. If you shorten the distance between the two balls in half, they will be pulled toward each other with a force *exactly twice* the force they had before you moved them.

Similarly, if you were to keep them in their original locations, but double the mass of each ball, the force of their attraction would be exactly the same as it was when you moved them two times closer to each other. This is because distance and mass work equivalently when it comes to gravity.

This is what Einstein discovered, and what changed so much of how we think about the way that gravity works.

Okay, okay, I know what you're thinking: That's all wonderful, but what does all of this have to do with Fortnite? Yes, obviously there is gravity on the island in Fortnite, otherwise everybody would just be floating around all the time.

As in any video game, gravity in Fortnite is very important. The game needs to be programmed in such a way to make sure that the players and the vehicles and other objects don't just fly off into the air, but gravity plays a much bigger role in the game than just that. Gravity controls how pretty much every object in the game (players, vehicles, materials, bullets) interacts with the player, the environment, and with other objects.

What makes you hurtle down toward the island when you jump off the Battle Bus? What makes you fall to your death when someone shoots the floor out from under you? What makes your Driftboard glide so effortlessly down the side of the volcano? Gravity, gravity, gravity.

Sort of. Of course, Fortnite is a video game, so there is no actual gravity on the island. Instead, the physics of the game have been programmed to imitate gravity.

But how does gravity in Fortnite work?

Epic Games is not spending hundreds of thousands of hours creating a one-to-one simulation of the Earth in order to create enough mass for the island to have its own digital kind of gravity. But efficient use of time is not the only reason Epic does not make gravity in Fortnite an exact simulation of gravity on Earth. If you made gravity in Fortnite the same as gravity on Earth, the game really wouldn't be very much fun to play at all.

Usually, people tend to assume that the gravity in a video game would be less than gravity on Earth, because it is seemingly so easy for characters to jump higher than real humans in the real world can jump, run faster than real humans can run, etc. However, this is not the case. In fact, if the force of gravity in Fortnite were less than the force of gravity on Earth (or even the same, for that matter), many things in the game would go very, very slowly.

The force of gravity on Earth is exactly 9.807 m/s². This means that, in a vacuum, an object dropped at the surface of the Earth will accelerate at 9.807 meters per second per second. If the gravity in Fortnite were the same as gravity on Earth, and you were to build a tower that was exactly 100 meters tall and then step off the side, it would take you approximately ten seconds to fall to the ground.

Now if you think about that for a second, you'll realize that it definitely does not take that long to fall 100 meters in Fortnite. So how long does it take? Well, based on my own brief experimentation, a fall from 100 meters takes approximately 3.5 seconds: much quicker than it would on Earth.

When you calculate gravity with the observations I made during my experiment in Fortnite, you come up with a gravitational force of twenty-eight meters per second. This is nearly *three times as strong* as the force of gravity on earth!

The reason for doing this is to give the game a nice snappy feel that we have come to know and love. However, if all they did was increase the force of gravity, and keep everything else about the game's physics equal to Earth, the last word you would use to describe Fortnite would be "snappy." That's because an actual human being that got dropped onto an island with 2.8 times as much gravity as Earth would barely be able to stand up, much less run around fighting off enemies and busting up buildings.

For example, the ledge just beside Paradise Park (the one overlooking the giant burger statue) is approximately two meters tall. If you have ever been to the ledge on your way into Paradise Park, you would know that it is also quite easy to jump up on. However, the world record for the highest jump by a human is only three meters. While the *Guinness Book of World Records* does not explicitly state as much, it is quite likely that this jump was, in fact, made right here on Earth. That means that you would need to be nearly twice as strong as the highest jumper in the history of humanity to make this tiny, easy jump outside of Paradise Park.

Similarly, think about some of the vehicles in Fortnite. By placing a waypoint marker at one end of the football field in Pleasant Park, while standing on the other side of the football field, one can confirm that it is a standard 100-yard American football field. Then, one can easily time how long it takes for a Quadcrasher to make it from one end of the football field to the other. That time is exactly four seconds. Quick math tells us that this means the Quadcrasher goes about 90 mph. Not only is this far faster than any quad that's ever been sold on earth (at least commercially), but remember: with the gravity on the Fortnite island, this same quad would be able to travel more than 250 miles per hour on Earth!

So what gives? How is it that all of these things can happen in such strong gravity?

The short answer is that they can't. In order to make a game like Fortnite actually fun, the developers have to simultaneously increase the force at which gravity pulls you toward the ground while also vastly increasing the strength of everything that needs to work against that gravity, far beyond what is possible in the real world.

In fact, you could probably argue that the most blatant way Fortnite breaks the laws of physics is in the way it defies the laws of gravity. Many of Fortnite's gravity-defying ways will be given extensive coverage in their own chapters later in this book. For example, we will discuss in detail how the Battle Bus manages to defy gravity to fly, what kind of technology allows the Driftboard to hover so effortlessly off the ground, and even what might be going on with those gravity-defying jetpacks.

But for now, it is enough that you know how gravity works in the real world, and how it is both stronger in the Fortnite universe and, for some reason, capable of being quite easily broken.

JETPACKS

Jetpacks have to be one of the most iconic science-fiction inventions of all time. For nearly a century, people have dreamed about strapping some kind of technological device onto their back and being able to fly, not quite like a bird or a plane (which, after all, both need wings to fly), but rather like Superman himself: darting around the clouds without any hindrances at all, as if the laws of gravity simply cease to exist. Given the long and storied history of these science-fiction wonders, it should come as no surprise that, from time to time, Epic decides to throw a few jetpacks onto the Fortnite island.

In Fortnite, the jetpack is a limited-time, legendary backpack item that was introduced in Season 4. From the moment you pick it up, the jetpack is equipped with a limited supply of non-replenishable fuel—when the fuel runs out, the pack can be discarded for the rest of the match. It's important to note that the jetpack can only be used for quick bursts, which last only a few seconds. If you keep the jetpack running for too long, it will overheat and stop working, possibly even when you are far up in the air, sending you plummeting to your death. For this reason, it's a good idea to always have a glider or two on hand when using a jetpack.

The jetpack can be very useful in Fortnite, and even with its limited range, it can greatly increase a player's mobility. It allows players to jump easily over relatively short ravines, or from the top of one building to another, or even just to get a quick boost of height to look around the environment and see where your enemies might be hiding.

The Fortnite jetpack is a rather small device. It has a metallic body, with two cylinders that appear to be engines of some kind on either side of another cylinder, which is presumably where most of the mechanical parts of the jetpack are located. At the top of the device is a red canister with a flammable label, indicating that it must run on some kind of flammable fuel.

So what exactly is a jetpack, and how does it work?

Before I can start to explain that, we have to make a clear distinction between two very similar things: a jetpack and a rocket-pack.

Believe it or not, the main difference between a jetpack and a rocket-pack is that a jetpack gets its thrust from *jets*, while a rocket-pack uses *rockets*. I know that sounds pretty complicated, but I'll try to break it down for you.

First: What is the difference between a jet and a rocket?

In order to answer that question, we need to go back to Sir Isaac Newton. Specifically, Newton's three laws of motion, which are responsible for the physics behind all jet engines, rockets, and certainly anything that anyone has ever called a jetpack.

Newton's first law of motion states that every object in a state of uniform motion will remain in that state of motion unless an external force acts on it. This basically means that all objects will resist changes to their motion (or lack of motion), and therefore must be pushed or pulled or otherwise acted upon with an outside force in order to be moved.

Newton's second law of motion states that force equals mass times acceleration. The important thing to remember about this law is that in order to overcome a force like gravity, the force that the jetpack exerts against the ground has to be greater than the force with which gravity is pulling back toward the ground.

Finally, there is Newton's third law of motion, which is the most important for this particular discussion.

Newton's famous third law of motion is this: every action has an equal and opposite reaction. This is the fundamental concept behind both rockets and jets. To visualize Newton's third law of motion, I'd like you to imagine that you are sitting in a rowboat with no paddles and your friend is sitting in an identical boat right next to you. Now, let's say your friend smells really bad, so you want to get as far away from them as possible. How would you do it? Without any paddles, the best you can do is to stick a hand or a foot out of your boat and push on your friend's boat as hard as you can.

What exactly happens when you push on your friend's boat? Do you move away from them? Do they move away from you? Or do you both move? Well, if you've ever spent any time in a rowboat, you'll know right off that you will both move if you push on the other boat—your boat and the other boat will move away from each other. This is because of Newton's third law of motion: you cannot push the other boat away without having an equal and opposite effect on your own boat.

JETS

Okay, now that we've gone through some of the underlying scientific concepts, it's time to look at how a jetpack might actually work. To do that, we need to start by looking at the jet engine.

The jet engine was certainly one of the greatest inventions in modern aviation, allowing airplanes to travel much faster, with much greater efficiency, than they ever could before. While there are many different kinds of jet engines, they all work in essentially the same way. A large turbine (which is like a really powerful fan) brings air into the jet engine. The air is then mixed with some sort of combustible jet fuel, which is then ignited inside a chamber. When this air and fuel mixture is ignited, it undergoes a chemical transformation that makes it get very hot while simultaneously expanding enormously in volume. What was once a relatively small volume of room-temperature air and jet fuel instantly becomes a very large volume of hot exhaust.

All of this exhaust has to go somewhere. If the compartment where the fuel and air ignite were completely enclosed, the entire engine would just blow up, sending shards of metal and engine parts everywhere. Thankfully, most jet engineers know about this whole explosion thing, so they like to make a small opening at the back of this chamber for the exhaust to escape from. As the exhaust escapes from this chamber, it travels into a compressor, which is essentially a series of tubes that get smaller and smaller as they get further away from the ignition chamber.

To understand what this process does, just think about what happens if you take a regular garden hose and stick your thumb over the end of it to make the hole that the water is coming out of smaller. What happens? The water, being under the same pressure as it was before you put your thumb there, now has to move a lot faster to get out of the smaller opening. The same thing happens inside the jet engine. As the expanding gas travels through the smaller and smaller tubes of the compressor, it goes at a faster and faster rate of speed until it is eventually let out through the jet nozzle at the back of the engine. On a typical jet plane, the jet engine is mounted on the wing with the exhaust nozzle pointed toward the back of the plane. Now we come back to Newton's third law of motion. If every action has an equal and opposite reaction, then it must be true that the action of these expanding gases exiting at an extremely high rate of speed toward the back of the plane will push the plane forward. This is called thrust, and it is a measure of how much force is being used to push something forward in this manner.

Okay, so what about rockets?

For the most part, rockets work very similarly to jet engines. In both cases, you have a combustible fuel that gets mixed with oxygen and ignited so that the gases expand and exhaust in the opposite direction that the rocket is intended to go, thus using Newton's third law of motion to propel the rocket forward.

The main difference between a jet and a rocket comes from where each type of engine gets its oxygen. We know that all fuel needs oxygen to burn. However, rockets were largely designed to fly into the outer reaches of the atmosphere, and even into space, where there is no oxygen just hanging around outside the engines. So, while a jet draws air in from an opening at the front of the engine in order to mix it with the fuel, a rocket has to carry its own oxygen. Usually, a rocket's oxygen will be carried in a separate tank and mixed inside the chamber where the incineration happens. Not only does this allow the rocket to function in environments where there is no oxygen, it also means that there is no need for the rocket to suck in a large quantity of air at a high rate of speed.

As a side note, if you have ever seen a video of what happens when a bird flies in front of a jet engine, you will understand why this is a big advantage. Placing anything in front of a jet engine is like placing it in front of a giant, deadly vacuum cleaner: it will instantly get sucked into the blades, probably destroying both the engine and whatever has the misfortune of getting sucked into it. Furthermore, it is easy to see why having this particular part of a jet engine right next to, say, somebody's nice long ponytail could be problematic. With the rocket, there is no need for that.

In any case, the fact that rockets need to carry their own oxygen means that they need to be significantly heavier than jets, which in turn means that they need to carry even more fuel and hence even more weight.

Throughout the history of humanity's many attempts at creating a viable, working jetpack, both rocket and jet engines have been used. Even though the jetpack in Fortnite has the word "jet" right in the name, there is ample reason for thinking that the pack could actually use rockets, so we will discuss both kinds here.

THE PROBLEM WITH JETPACKS

If Newton's third law of motion is responsible for making jetpacks and rocket-packs fly, then his first and second laws of motion are to blame for keeping so many jetpacks firmly rooted to the ground.

According to the first two laws, an object at rest will stay at rest unless acted on by another force, and that force, in order to fly, must be greater than the force of gravity keeping the jetpack on the ground.

Rockets and jet engines each overcome these two laws in slightly different ways. The jet engine uses forward thrust to propel a plane forward, but not to push it up. The upward motion comes from the lift generated by the shape of the airplane itself (for more discussion of lift, see the chapter on Skydiving and Gliding). Essentially, this means that the airplane only has to generate enough thrust to move air over the wings at a high enough rate of speed to create lift. A rocket, on the other hand, does not generate lift with wings as an airplane does. It must rely solely on the *power of the rocket engine itself* to overcome inertia and gravity.

I don't want to make any untoward assumptions, but I bet that if you take a good long look down at your own body, you'll notice that you, like a rocket, do not have any wings. Thus, getting a person off the ground is much more akin to getting a rocket off the ground than an airplane.

Unfortunately, as with a rocket, this means that a jetpack will need to generate a great deal of thrust to overcome the force of gravity and get both itself and its pilot into the air. However, in order to generate this much thrust, the rocket will need a rather large quantity of fuel (which means the rocket-pack will weigh more due to the increased fuel, which means it will need even more thrust, which means it will need even more fuel, etc.).

This is where the real problem with jetpacks comes in: it is very difficult, with something that can fit on the human body, to generate enough power to overcome Newton's first two laws of motion, at least for any significant length of time. Anyone hoping to make a working jetpack will have to figure out a way to create a large amount of thrust, with a very small amount of weight, in a package that can fit on a human body. This is a problem that no one has satisfactorily solved yet.

These problems are not the only issues with jetpacks. Another problem is that of maneuverability. As we discussed earlier, the vast majority of humans don't have wings and are not designed to be particularly aerodynamic (just take a walk around your local beach on a hot day if you don't believe me). Think about it this way: if you were to take an airplane up to 50,000 feet and drop it without turning on the engines, there is a pretty good chance, especially under ideal conditions, that a good pilot could land that plane somewhere (as the famous Captain Chesley "Sully" Sullenberger did on the Hudson River in 2009). This is because the shape of the airplane (aerodynamic body, wings, etc.) will allow it to still maintain some degree of lift as it accelerates toward the Earth (again, see the chapter on gliding for a more detailed discussion of this). A human (or even a giant lizard, tomatohead, or banana), on the other hand, without a working jetpack, will not be able to simply dive for a little while and then pull up and glide, because there is nothing on the human body with which to generate lift.

Of course, you could simply attach wings to somebody in addition to adding a jet or rocket-pack, but then you really don't have a jetpack, do you? I would call wings with a jet engine attached "a plane" (or at least some kind of a wing-suit, or something). Thus, any design for a functional jetpack will have to include with it some ability to maneuver around, most likely by changing the direction of the thrust as it is coming out the back of the rockets or jets: one more obstacle to overcome before you can have a real jetpack.

Yet another problem is the effect many designs of the jetpack would have on the person wearing them. Jet engines and rocket engines both burn fuel at extremely high temperatures and both vent super-heated exhaust that is usually over 2,500°F. As you can imagine, it would be very painful (or even fatal) to have any part of your body come in contact with this exhaust. Of course, you also have the problem in which a jet engine will generate a great amount of suction at the top of it, against which no item of clothing or hair would be invulnerable.

A BRIEF HISTORY OF JETPACKS

Unlike many of the items found in Fortnite and discussed this book, there have actually been a great number of jetpacks invented, with varying degrees of success. Before we get into those, let's look at where this whole idea came from in the first place.

There have been stories about people using jetpack-like devices going back a very long time, with some reports even claiming that monks in ancient China strapped rockets to themselves in order to fly up into the sky. The first jetpack in popular culture, however, appeared in *Amazing Stories*, which is a magazine that has published short novels and stories of science fiction and fantasy since the early twentieth century. The cover of the August 1928 issue features a picture of a man with what looks like a jetpack strapped to his back. He seems to be holding some kind of a controller tethered to a belt, as he flies around in the air. This rocket-man was the primary illustration for a story called *The Skylark of Space* by E. E. "Doc" Smith. It should come as no surprise that in this story, the invention of a working jetpack was preceded by the discovery of a new element that provided an enormous amount of power with a very small amount of weight.

A little more than twenty years after this appearance in *Amazing Stories*, jetpacking saw its first great rise in popularity with the 1949 film *King of the Rocket Men*, in which a scientist creates a "sonic-powered rocket backpack" (don't even ask me how that's supposed to work) that shoots fiery streams of smoke and fire behind it, looking very much like our modern conception of a jetpack.

Since that time, there have been many depictions of jetpacks in popular culture and science fiction, with nearly as many real-life attempts happening alongside them.

For example, there was the de Lackner HZ-1 Aerocycle in 1955, which was created for the US Army as a one-man personal helicopter. It was basically a small platform on top of a number of rotating helicopter blades and some pontoons (it sort of looked like a huge version of a modern drone with a podium stuck on top). It did not have a rocket or a jet engine of any kind, and by all accounts was nothing short of a deathtrap.

There was Wernher von Braun's 1949 Jet Vest, which used rockets to allow a soldier to jump over tall obstacles, as well as Project Grasshopper, a simplistic rocket belt commissioned by the US Army in 1958: both projects that never quite worked for more than a short jump (and which posed great risk to their pilots and great cost to their financiers) and were all abandoned before too long.

The first real, working jetpack (well, sort of) was invented in 1961 by a company called Bell Aerosystems, and was called the Bell Rocket Belt. The first such device to achieve anything like real flight, this was decidedly a rocket, in that it carried all of its own fuel and didn't need to suck in any oxygen from its environment. The Bell Rocket Belt actually worked by putting hydrogen peroxide into contact with silver, creating a violent chemical reaction that expanded the compressed liquid hydrogen peroxide into a gas almost instantaneously, directing the thrust this created out small nozzles on either side of the back of the rocket belts. The pilot of the Rocket Belt could adjust the direction that the exhaust flowed through these nozzles, which allowed him to maneuver without the use of wings or other aerodynamic apparatus. This was still a significantly dangerous design, and while it did technically work, they never got it to run for more than twenty seconds without needing to be shut down and refueled.

Since 1961, there have been many other similar types of inventions, using different combinations of fuel, aeronautic design, and thrust to try to solve the same basic problem, but none of them wound up working much better than the Bell Rocket Belt.

The most successful of these subsequent jetpack designs was created by a man named Richard Browning, who was recently awarded with the *Guinness Book of World Records* record for the fastest speed flown using a jetpack (actually, the record is called "flying in a personal flight suit"). To win this record, Browning flew his invention, which he calls the Daedalus Flight Suit, at a blistering 32 mph. The Daedalus Flight Suit is very much a real jetpack, as its engines are actual jets that take in air and use kerosene as a liquid fuel. Browning's real innovation with the Daedalus Flight Suit is his ingenious addition of thrusters that attach to the end of the arms. This allows the jetpacker a much greater degree of maneuverability and successfully mitigates one of the greatest challenges for jetpack inventors. All that said, even the Daedalus Flight Suit didn't manage to make much progress with the fuel-to-weight-ratio problem, as it only has a maximum flight time of about ten minutes.

What does all this mean for the Fortnite jetpack? Can you actually make one like that in real life?

Probably not. The major problem with the design of the Fortnite jetpack is that it just does not seem to have any type of maneuverability whatsoever. We can see quite clearly from even a brief inspection of the device that the exhaust is vented straight out the bottom of the jetpack, allowing for thrust in only one direction. This means that, without any type of wings or wing suit, your Fortnite jetpack would have no means of maneuvering through the air, and you would probably crash straight into the ground as soon as you took off.

But where does the Fortnite jetpack seem to get it right? Well, it is possible that it is, in fact, a real jetpack (as opposed to a rocket-pack), although it is not quite clear where the air intake could be. There are a few vent-looking things on the back of the jetpack, but they are not positioned in such a way as to optimally suck in air (which would be at the top of the jetpack, as it is with the Daedalus). The tops of the engines on the Fortnite jetpack have two hoses that run directly to what is clearly the fuel canister at the top. Also, there is the issue of the big cylinder in the middle. What is that? It almost looks like a carburetor, although that would be a very inefficient way of mixing air with fuel for something like a jetpack.

A better guess is that this is actually a rocket-pack of some kind, which uses a small amount of some yet undiscovered combustible fuel and some kind of fuel cell inside that center cylinder to create enough thrust to lift someone off of the ground. Still, though, without a clear means for steering the jetpack, we're going to have to say that this is probably not a plausible device at all.

DRIFTBOARDS

The year was 1989. George H. W. Bush was president, the Electric Slide was tearing up dance floors around the world, and Sir Tim Berners-Lee had just finished inventing something called a "hypertext markup language" that would allow people to create "pages" on a "World Wide Web."

For my part, I was nine years old, living on Long Island, and getting regularly teased for being the only boy in my class whose mom still made him wear penny loafers to school. Of course, the most important thing that happened in 1989 was the release of *Back to the Future Part II*, which was at the time (and perhaps is still) the only sequel to ever truly live up to its predecessor.

What was so special about this movie? Well, apart from the fact that Biff becoming the mayor of Hill Valley so closely predicted our current political situation in 2019, it was a little pink toy, stolen by the protagonist from a little girl in one of the opening scenes, that would forever change history.

I am talking, of course, about the hoverboard. In *Back to the Future Part II*, Michael J. Fox's character's hoverboard was a hot pink, Dogtown–style skateboard deck that hovered about a foot off the ground, seemingly without the use of fans, turbines, jets, rockets, or any other type of thrust-creating engine.

This original hoverboard seemed to function much like a standard skateboard, but with less friction and the ability to jump a bit farther. Not only was this just a really freaking cool scene in a movie that instantly captured the imaginations of every kid (and most adults) who saw it, but the director, Robert Zemeckis, actually said in an interview that the technology to make a hoverboard was totally real, but all of the toy companies were being kept from releasing it by concerned parents who feared that it was too dangerous for their children to own.

Of course, this claim turned out to be a hoax. There were no hoverboards like the one in the movie back then, and there are no hoverboards quite like it now. This was far from the first hoverboard hoax, with many weird and wonderful videos of supposedly real hoverboards being circulated widely in the years since then. Nonetheless, the hoverboard from *Back to the Future Part II* captured the world's imagination in a way that few other fictional inventions ever had. Because of this, many people would spend years trying to make a hoverboard that really does work like the one in the movie, though with greatly varying degrees of success.

THE FORTNITE DRIFTBOARD

Okay, yes, I know. Technically, the hoverboard that most people think of when they think of Fortnite is actually called a Driftboard. And yes, I know that there was something called a hoverboard in *Fortnite Save the World*, but for this chapter, we're going to focus on the Driftboard from *Battle Royale*.

The Driftboard, on first glance, seems to be quite similar to the hoverboard from *Back to the Future Part II*. They are both more or less flat boards with no obvious engines attached to them, or audible propellers of any kind, that seem to hover about a foot off the ground and can be ridden more or less like a standard skateboard. The main difference, on first inspection, is simply the size: the Driftboard in Fortnite is roughly the size of a snowboard, not a skateboard. Like the *Back to the Future Part II* hoverboard, however, it does seem to be slightly thicker than its non-hovering counterpart.

The Fortnite Driftboard has two circles on top of its deck and two circles on the bottom. The two circles on the top are where the rider's feet go. There are no bindings, as there would be on a snowboard, but the circles seem to have the ability to stick the feet of the rider. There is no obvious mechanism for this sticking action, so I'm going to go ahead and guess that it's either magnets or chewed-up bubblegum.

The two circles beneath the Driftboard seem to be the source of its hovering ability. When the board is in use, one can see a sort of hazy blue distortion wave coming out of these blue circles and expanding toward the ground.

Of course, there's also the fact that the Fortnite Driftboard has a booster engine of some kind that allows the rider to get a few seconds at a time of very fast forward thrust before the booster has to recharge itself. A bar of light on the back of the Fortnite Driftboard glows green when the booster is ready to go and then seems to shoot out some kind of high-powered exhaust while the booster is functioning, before turning orange while the power is replenishing.

The Driftboard has always been my personal favorite means of transportation around the Fortnite island. It is quite fast, allowing you to traverse the entire island in just a few minutes. It also has the added benefit of protecting the rider from fall damage, presumably because the hover effect somehow dampens the fall just before impact.

What is the science behind the Fortnite Driftboard? Is it based on anything that is really possible with today's technology? What are the fundamental scientific concepts that either allow or disallow such an invention to work?

Unfortunately, you probably know already that nothing like the Fortnite Driftboard is readily available at your local skate shop (I am, of course, not counting those two-wheeled handlebar-free Segway-like things that people call hoverboards, because, you know, they don't actually hover), and while there are a few hoverboard technologies that accomplish some of the things found in both *Back to the Future* and Fortnite, nothing has yet been invented that does all of it. The hoverboards that do kind of, sort of work rely on two related, but decidedly different, technologies to achieve their levitation. Both of these technologies rely on one basic, scientific concept: magnetism.

WHAT ARE MAGNETS?

Now I don't know about you, but magnets are pretty much the closest thing to magic that I've ever seen with my own two eyes. They have the ability to move things almost as if they were Luke Skywalker using the Force, with nothing seen or felt between the two objects being pushed and pulled. But how does that happen? How does one object exert a force on another object without touching it or seeming to come in contact at all?

The answer to this is magnetism. Humans began to learn about magnetism in ancient China, when they first realized that a magnetized needle floating in water would always point in the same direction, no matter where the magnetized needle was. This was, of course, the first compass, and as you probably know, a compass is just a small magnet that points toward the very tip of a very large magnet called the Earth.

But why? What is the force that actually makes that needle point north?

To answer this question we need to get down to the very, very, very, very small things that make up the universe: atoms and their even smaller component parts, electrons.

The easiest way to explain how magnetism works is to start by picturing an atom in your mind. Persnickety physicists will tell you that the classic illustration of the atom (you know the one: a little cluster of protons and neutrons at the center and lots of little electrons spitting around it like planets orbiting a bumpy little star) is terribly inaccurate. While they would technically be correct (quantum phenomena being very difficult to represent in two-dimensional images), for our purposes, it works just fine.

Okay, so you have your little electrons spinning around the neutrons and protons like tiny little planets, and also like many planets, these little electrons are not only circling the nucleus of the atom, each one is also spinning on its own axis, similar to the way the Earth spins as it moves around the sun.

Now here's the really important part: it is this spinning—the spinning of the electrons themselves, and in particular which direction they are spinning—that creates the magnetic field. In fact, each and every little electron in the universe is constantly generating its own small magnetic field, with each one either going north or south depending on the direction that the electron is spinning. With me so far? Good.

So, in the vast majority of elements, electrons come in pairs, with each individual electron in each pair spinning in a different direction. Because each pair is simultaneously generating both a north-facing magnetic field and a south-facing magnetic field, the sum total of the magnetic field generated by each pair of electrons is zero. This is kind of like trying to walk up the down-side of an escalator: both things are in motion, but the motions of each cancel the other out. At least in most cases.

There are, however, certain atoms that work a little differently. There are certain types of atoms whose electrons do not all travel in pairs. These are called unpaired electrons. With most of these atoms, the magnetic field generated by each unpaired electron is still canceled out by another unpaired electron spinning in a different direction somewhere around the same atom. However, there are three specific elements that have unpaired electrons with spins that are not canceled out in each atom. These elements are iron, cobalt, and nickel.

What do you think happens when an atom has unpaired electrons whose spin is not canceled out? Well, obviously you're going to have an entire atom with a net magnetic pull equal to that of its one unpaired, uncanceled-out electron.

This is why some elements, like those mentioned above, can be magnetic, while others cannot—when each atom has one electron spinning up one small magnetic field, the sum total of all these little magnetic fields is a much larger magnetic field. These are called ferromagnetic elements.

That said, even though all ferrous metals *can* be magnetic, they are not all actually magnetic all the time. So what happens? What makes one piece of iron magnetized, and another not magnetized?

To answer that question, we need to zoom out to a slightly larger scale than the individual atom. Instead of picturing a single atom, I'd like you to picture a whole bunch of atoms just hanging out near each other, just sort of floating around, being atoms. For this example, we'll say that they are iron atoms. Now, because iron is a ferromagnetic element, each one of these atoms has a magnetic charge of either up or down. Left to their own devices, most of these atoms will find other atoms with the same charge to hang out next to, clustering in groups of atoms with the same magnetic charge. These clusters of ferromagnetic atoms are called *domains.* Each domain, being made up solely of atoms with the same magnetic charge, will itself have a particular magnetic charge, with all of the poles facing in the same direction. This uniformity only extends to the individual domain, however, and even a tiny piece of iron is made up of millions of such domains, with each one facing the different direction. On the whole, with such a large number of domains facing in different directions, the magnetic charge will be canceled out and the piece of iron as a whole will be magnetically neutral.

However, if you can get all of the domains to face in the exact same direction (which can happen for a few different reasons that I will discuss below), you wind up with a magnet that is actually capable of extending a magnetic field beyond its own physical boundaries.

So how do you get all of these domains to face the same direction? The simplest method is to just use another magnet. When you take a strong magnet and touch it to an unmagnetized piece of iron (in other words, a piece of iron with all of its domains facing different directions), the strong magnet will pull all of the domains from that piece of iron into line with its own magnetic field.

Another way that you can pull all of these domains into the same direction is with electricity. Electricity and magnetism are very much connected, and can often be thought of as two sides of the same coin, in very much the same way that space and time are two sides of the same coin. When you run electricity through a magnetic material, it creates a magnetic field, pulling all of the poles of the domains into the same direction around it.

To quickly review: each atom in a magnetic material, as a result of the unbalanced spin of its electrons spinning, creates a pole in a certain direction. A bunch of atoms all pulling in the same direction right next to each other is called a domain. Usually, all of the domains inside a magnetic material are pulling in different directions, making the material as a whole cancel out its own magnetic field. But when all of the domains are pulling in the same direction, and therefore not canceling each other out, they create a magnetic field that pulls beyond the reach of the physical element itself.

And what does this field look like? Well, the field pushes out of the north side of the magnet, sweeps around, and then comes back into the south side of the magnet, pulling anything on that side that is also magnetic. That's why if you stick the south end of a magnet onto a refrigerator it will stick, but the north end won't.

That is also why you can stick the north end of a magnet to the south end of another magnet, but if you try to place both north ends together they will push each other away. It is this pushing force that is the basis for any hope we have of magnetic hoverboard technology.

BACK TO THE HOVERBOARDS

As I said earlier, there are basically two types of hoverboards that exist in the world today, and each of them uses magnetism in different ways. The first kind uses something called magnetic levitation, while the other uses a phenomenon called flux pinning.

Let's start with the basic idea of a magnetic levitation board. If you've ever played with two magnets, you know that depending on which way you turn them, they will either be pulled together or repelled apart. This has to do with the direction of the pole of the individual magnets. When two north poles point at each other, they repel each other, and when a north pole points at a south pole, they attract each other.

Imagine that you could take one magnet and shape it into a skateboard deck with its north pole facing down, and then take another magnet and shape it into a giant field with its north pole facing up. What would happen when you put the skateboard on the field? Exactly: they would repel each other, and the skateboard would hover.

Well, sort of. In truth, the skateboard would probably hover for about a half a second before flipping over and smashing into the field, but you get the idea: one can, theoretically, use the natural repelling effect of two magnets to make something (like a board) hover. Of course, it is not actually possible to build a hoverboard like this, as you'd never get a large enough magnet to make your field, and even if you did, you would never be able to make it stable enough to ride.

So how do you get a magnetic hoverboard to actually, you know, hover? The science behind the first type of hoverboard we're going to discuss is pretty darn cool (though perhaps not quite cool enough to make an actual Fortnite Driftboard work, alas). This type of hoverboard, popularized by the car company Lexus in a marketing campaign in 2015, uses a quantum phenomenon known as *flux pinning* to achieve its levitation.

Flux pinning takes advantage of a unique property of superconducting materials. A superconductor is a type of material that can conduct electricity without any resistance. This means that no energy is lost when passing electricity through the material, and no heat is generated either. Now, as it turns out, all superconductors have the interesting property of expelling magnetic fields. This means that instead of a magnetic field passing through a superconductor (the way it would pass through pretty much anything else), the field actually gets bent around the superconducting material.

There are, however, a few superconductors that have the additional property of simultaneously expelling magnetic fields while also allowing small vortexes of magnetic flux to pass through them. What happens then is that the magnetic field mostly passes around the superconductor with little pins of it pulling through the superconductor. The effect of this is that the superconductor will become locked in place in its relation to a magnet when placed beside one. So if you take a piece of this kind of superconductor and place it over a stationary permanent magnet, even an inch or two, it will remain levitating in place there. This is called quantum locking or flux pinning.

Let's say we take a few of these superconductors and put them into a skateboard deck. Then we take that skateboard deck and we put it on top of a magnet. What will happen? That's right, the skateboard will hover. If you then make a long magnetic track instead of a single magnet, you can run the hoverboard over that track, which is exactly what Lexus did in its 2015 commercial.

There are, of course, a few problems with this design. One problem is that this kind of hoverboard will only work over a predetermined track. You can't just run it around anywhere. The other problem—and this is really the big one—is that superconductors like this have to be kept extremely cold: –183°C, to be precise.

Also, in order to get them to hover, you have to actually place the hoverboard above the track while the superconducting material is warm, and then use liquid nitrogen to make it cold enough to attain its superconductive properties. It is only at this point that you get the quantum locking that provides your levitation. This means that you need to reset the height of your hoverboard with liquid nitrogen each and every time you use it—clearly not a good option when you need to hop on your board to get away from a human-sized banana with a machine gun.

There is also the impracticality of having to keep the superconductor so cold. In the real world, you would need to constantly be recharging it with new liquid nitrogen every few minutes. But still, even if the mad geniuses at Vindertech figured out how to make a superconductor work continuously in a snowboard deck, you would still need to have the rails to make this one work.

This brings us to the second—and for our purposes most promising— hoverboard technology.

Before we can get to how this hoverboard works, I need to explain a bit more about electricity. Whenever a magnetic material is subjected to changes in magnetic polarity (like passing magnets pointed in different directions right past it over and over again), an electric current is created in the material. This is the basic technology behind all electricity generation. In an electric generator, large magnets with alternating polarities are placed beside each other on a wheel inside a circle of copper wire and then spun at very high speeds, constantly passing the different poles past each section of wire to create an electric current.

This second type of hoverboard uses a phenomenon known as Lenz's Law to create a downward-facing magnetic field. How this happens is a bit complicated, but it basically involves an electromagnetic engine spinning magnets around fast enough to generate electricity in a wire in such a way as to create slightly more downward-facing magnetism than upward-facing.

If this downward magnetic field were put over a material that is magnetic, nothing would happen—the hoverboard would just stick to the material. However, remember that changes to magnetic fields create inductive electricity in nonferrous materials. So when this hoverboard is placed over a sheet of nonferrous material (like copper, for example), the net downward magnetic flux produces a small electric charge in the copper just beneath it. This electric charge, in turn, creates its own magnetic field that pushes upward. Thus you have magnets inside the board being spun around fast enough to create a magnetic field pushing down on the copper flooring; this induces a small electric charge in the copper flooring, which, in turn creates its own magnetic field pushing upward against the board. Thus you have a magnetic field in the hoverboard pushing down and a magnetic field in the material beneath pushing up. This gives you levitation, and this is how you get yourself a kind-of sort-of working hoverboard.

Now for the bad news: first of all, this hoverboard relies on having a decent amount of power to spin the electromagnetic engines in the board, so it doesn't last very long. Also, and more importantly, it only works over very specific materials. So unless you cover the entire Fortnite island with a thin sheet of copper, the Driftboard could not function exactly like this one.

Now, as great as it is that these inventors manage to get as far as they did with hoverboards, can either of these technologies really be used to explain the Driftboard in Fortnite?

I think we can safely say that there is no possible way that the Fortnite Driftboard uses any type of flux pinning, as it would need to be kept so very, very cold, and the hover height would have to be reset every time you use the board, which clearly is not happening (and which would not be very practical, in any case).

But what about the other kind of hoverboard? That really depends on the material that the ground is made out of. As of right now, the only way to get this kind of hoverboard to work is with a perfectly flat copper plane on which to ride. One can use other nonferrous materials than copper (like silver, gold, aluminum, etc.), but it still seems unlikely that the entire Fortnite island is made out of any these just to facilitate the Driftboards.

SKYDIVING AND GLIDING

Without a doubt, skydiving and gliding are some of the most important game mechanics in Fortnite. Together, they are the first actions you take when you step off the Battle Bus (except for thanking the bus driver, of course, which *I'm sure you all do every single time*) and try to pick a place to start your search for all that sweet, sweet loot. Not only that, skydiving and gliding are two of the key ways that a player gets around the Fortnite game map, and they can even be an important part of combat, especially when you get to the endgame.

So how exactly do skydiving and gliding work in Fortnite, and are they different from the way they work in the real world? Would gliders like the ones we see in the game actually function in real life?

To answer these questions, we first need to look at some of the scientific concepts that factor into skydiving and gliding in the real world.

To examine the basic physics of skydiving, we must begin by looking at the interaction of gravity and wind resistance, or drag. Gravity is the force that pulls us toward the Earth and keeps us from floating off into space. In the chapter on gravity, we discussed the deeper reasons why gravity does this, but that knowledge is not so necessary when considering the issue of skydiving and gliding. For these things, it is enough to understand gravity the way that Newton understood it: what goes up must come down. When someone jumps out of a plane and begins their free fall, gravity will pull them toward the Earth with a more or less consistent force for their entire way down.

And yet, if gravity remains consistent throughout the fall, how is it that skydivers can speed themselves up and slow themselves down during a free fall? Wind resistance.

But we're getting ahead of ourselves. Let's take all this one step at a time. You start by jumping out of the Battle Bus. What happens immediately upon your exit? Two forces start to work on you immediately: gravity and wind resistance. Gravity is pulling you down toward the Earth, so you will instantly start to accelerate, hurtling faster and faster toward the ground.

This is when things get interesting. If you had jumped from a large, blue, balloon-lofted bus that happened to be flying over Mars, for example, you would just keep on accelerating nearly all the way until you hit the ground. In fact, if you were skydiving over Mars, opening up a parachute or a glider wouldn't even slow you down enough for you to notice any kind of significant drop in your speed. This is because Mars, unlike Earth, has virtually no atmosphere (which is another way of saying that Mars has no air), and it's the atmosphere that slows you down.

So why does the air in the atmosphere slow you down? Mainly because, believe it or not, air is a fluid. This means that it is a substance without any specific, fixed shape, and that it yields to external pressures (such as, for example, a falling, human-sized banana). As you are diving through the air, you are actually pushing all of these tiny little air molecules aside, and while you're doing that, all of those tiny little air molecules are pushing right back at you (once again, we come upon Newton's third law of motion).

Think about it this way: What happens when you dive off a high diving board into a pool? When you first leave the board, you start to accelerate very fast, but when you hit the water, you instantly slow down. This happens because the water, being far denser than air, pushes back at you with greater pressure than the air does, slowing you down. The same thing would happen if you were to dive from outer space into Earth's atmosphere. At first you would accelerate very fast, but as soon as you got into the atmosphere, where the density of molecules is much greater than out in space, you would start to slow down (though, of course, the friction of doing so would make you burn up faster than falling into a pit of lava, but you get the point).

Now, I know what you're thinking—wouldn't it be awesome if there were no atmosphere on Earth so we could just skydive superfast all the way down? While yes, that would be really cool, it would actually take pretty much all of the fun out of skydiving, because it's actually this very wind resistance that, while it does slow you down, lets you maneuver around through the air, allowing you to reach distant parts of the map and giving you the sensation that you're actually flying.

Probably the first thing you notice after you jump out of the Battle Bus, but before you open your glider, is that you can manipulate your body in the air to move around. When you dip your right side down, you turn right. When you dip your left side down, you turn left. When you point your head at the ground, you go faster, and when you lay your body flat and parallel to the ground, you go slower. All of this happens because of wind resistance.

The simplest way to understand this is to think about the air as not being empty, but filled with billions and billions of tiny little molecules (which, of course, it actually is). As gravity is pulling you down toward the Earth, you are pushing away every molecule that happens to be floating directly between your body and the ground. When you spread your body out flat, there are far more molecules between your body and the ground than when you are diving headfirst. This is because when you are diving, you are only pushing aside the molecules that are directly on top of your head, which, even if your head is a giant tomato, is still a lot fewer molecules than when you are spread out flat and pushing away everything under your legs, feet, arms, torso, face, etc.

Regardless of how you decide to position your body while falling, eventually you will accelerate to a speed at which the force needed to push aside all of these molecules equals out with the force of gravity pulling you down toward the Earth. When you reach this speed, you will stop accelerating and remain at a constant speed until you either change your position, open some type of parachute or glider, or crash into the ground.

This speed, the speed at which you stop accelerating and start to drop at a constant rate, is called your *terminal velocity*.

Okay, are you still with me? Are you getting all this? Do you need a snack? Go ahead: get yourself a snack. I'll wait here.

Now, there are two main points I want you to keep in mind before we move on to the discussion of Fortnite gliders:

1. The more of the surface of your body is parallel to the ground, the greater the wind resistance will be pushing against you, and the slower you will fall.

2. No matter how much surface area you create between yourself and the ground, eventually you will reach an equilibrium called a terminal velocity, at which point you will stop accelerating and maintain a constant speed.

Now, keeping these things in mind, think about how a parachute works. When you open it up, you are creating this huge amount of surface area parallel to the ground, meaning there are now so many more air molecules you will need to push aside on your way to the ground. And as you now know, the more air molecules you push, the slower you will go, and the slower your terminal velocity will be. If you have a tiny little parachute, then you'll have a terminal velocity that's a tiny little bit slower than you were falling before you opened it. If you have a great big parachute, then your terminal velocity will be a lot slower than you were falling without a parachute. If you're really lucky, you can even make your terminal velocity slow enough that your body doesn't splat into a puddle of goo when you finally and inevitably make contact with the ground.

Okay, so that's what would happen when you open a parachute while skydiving, but we don't have parachutes in Fortnite. No, we have "gliders," which are not exactly the same, are they?

Well, they would function quite similarly to a parachute if all you were to do was open up your glider and fall, but that isn't going to get you where you want to go. A glider is different from a parachute mostly in that, instead of a canopy (what you call the actual cloth part of a parachute), you have some sort of wing, or foil.

Gliders, both in Fortnite and in real life, work based on similar principles as airplanes. Both gliders and airplanes achieve their ability to fly by generating lift with a foil, or a wing. The wings on both gliders and airplanes are angled slightly upward and are more curved on their tops than on the bottoms. This causes the air that flows over the top of the wing to travel a longer distance at a higher rate of speed than the air going underneath the wing, which in turn causes the air above the wing to becomes less dense (meaning that the air molecules are spread farther apart from each other) as it's spread out over the longer distance of the top of the wing's curved surface. Meanwhile, the higher-pressure air underneath the wing is being deflected away from the wing as a result of the wing's angle of attack. As a result of our old friend Sir Isaac Newton's third law of motion, this deflection pushes up on the wing, creating a force that pushes the airplane or glider off the ground. This force is called lift, and it's the basic principle behind all fixed-wing flight.

The key thing to remember about lift is that the air has to be traveling fast enough as it deflects off the wings to actually cause the plane (or glider) to gain altitude. In an airplane, this speed is accomplished with the use of engines. The engines on an airplane do not actually lift the plane up so much as they push the plane forward. It is this forward motion that increases the airflow under the wings and creates lift.

As you probably already know, there are no engines on a glider. So how does a glider create lift? With a glider, the speed is primarily created by gravity. As gravity pulls the glider down toward the Earth, the speed of the glider accelerates until it is going fast enough for the airflow under the wings to generate lift.

For this reason, in real-life gliders, one can maneuver into a dive to gain speed and then, once the glider is going fast enough, pull up and actually start to gain altitude until the force of gravity slows it down and it needs to dive again to gain speed. The gliders in Fortnite, for some reason, do not seem to have the ability to climb in altitude at all. I tried to think of some scientific reasons why it might not be possible for these kinds of gliders to fly up instead of only down, but I couldn't really come up with anything. In all probability, the gliders were simply programmed not to gain altitude so that players wouldn't have the ability to just fly around indefinitely.

While we're on the subject of gliders in Fortnite, we should probably talk about the standard design a little. As you can see in the game, the standard gliders have a small wing on each side of a square frame over which some kind of canopy has been strapped.

Clearly, the canopy at the top of the glider is meant to act similarly to a parachute. It increases the surface area of the glider, thereby increasing the number of molecules that must be pushed aside as the glider falls, thereby slowing the glider down (along with the person hanging on to the glider) to a more manageable terminal velocity.

The wings on the standard glider do not appear to be large enough to create a significant amount of lift, so what are they even there for? The most likely explanation is that the glider's wings are primarily included to aid in maneuverability. Unlike the canopy, the wings are fixed in their positions on the glider's frame, and therefore move in a precise and uniform way with the rest of the glider. This allows the glider pilot to have much more minute and granular control of the airflow over the glider, and hence more control over where the glider is going, how fast it is falling, etc.

I know what you're thinking. That all certainly may explain the gliders of the standard variety, but what about the ones that look like airplanes or helicopters or shards of glass or hamburgers? Well, in the opinion of this humble author, these marvelous gliders all fall under the scientific category of Totally Freaking Ridiculous.

One final note: no matter what design the glider is, every glider in Fortnite seems to have a very specific type of device connected to it that is notably similar to a real-life device used on some parachutes: an Automatic Activation Device (AAD).

A parachute AAD is designed to open the parachute when the skydiver reaches a certain distance from the ground. The idea behind this is that even if the skydiver forgets to open the parachute, or loses consciousness for some reason, the parachute will still open and the skydiver will not crash into the ground. Clearly, the gliders in Fortnite have a similar functionality: no matter how hard you try, you cannot skydive all the way to the ground (at least not from the Battle Bus).

Parachute AADs in the real world are set to open the parachute at a certain specific distance above the ground, which the altimeter determines based on a measurement of the air pressure. Before a skydiver takes off in a plane, they set AAD to detect their present altitude, then the device is automatically set to go off at 230 meters. Due to the fact that air pressure increases at a knowable rate the higher you get above sea level, the altimeter can be quite precise about when it opens the parachute. In Fortnite, however, this is clearly not the case. If it were, the glider would open at the same altitude, regardless of where the player is landing. If you have ever tried to glide over the volcano in Fortnite, you know that glider opens at a specified distance above the ground beneath it, and not a specified altitude over sea level, or some specific starting point. This means that if you are skydiving toward the top of a mountain and you are directly over the peak, the glider will open much sooner than it would if you were gliding over a lake or a valley or a particularly Dusty Divot of some kind.

From a technical standpoint, this probably means that the altimeter on the Fortnite gliders does not function based on air density at all. Instead, it must have some way of telling how far the glider is at any given point from the ground beneath it. This would be most easily accomplished with the use of some kind of sound or light sensor technology, similar to that used in crash-prevention technologies in cars.

This type of technology works by shooting light or sound at an object and then measuring the amount of time it takes for that light or sound to be reflected back to the sensor. Because light and sound travel at a knowable speed, the longer it takes for the light or sound waves to reach the sensor, the farther away the object is, and the shorter the amount of time it takes, the closer the object is. In this way, a car with this type of technology can tell when it's about to get into a crash and automatically apply the brakes. The gliders in Fortnite seem to have a sensor that works in a similar fashion, always being able to tell exactly how high above the ground the player is, no matter the elevation of the ground beneath them. When the player reaches a certain distance from the ground (which seems to be 100 meters, though some game modes it may be less), this sensor sends a signal to open the glider.

A quick practical note about how this type of technology affects gameplay: as a player, you should always be paying attention to the elevation and topography of the ground beneath you while you are skydiving, especially if it is your intention to free-fall for as long as possible before allowing your glider to open on its own. If this is your plan, you will need to keep yourself away from hills and mountains as much as possible, and try instead to free-fall over the lowest elevations you can. This will keep you free-falling longer and farther than you would otherwise. There is nothing worse than thinking you can free-fall pretty close to a sweet loot spot, only to accidentally fly over a hill and have your glider open much earlier than anticipated, often with the effect of allowing another player to reach that loot first (at which point they generally just kill you).

BATTLE BUS

Can you really lift a bus with a balloon?

Ah, the Battle Bus. For most players, this is one of the first things that makes you realize you are in a world with completely different rules of science and physics than the one you are used to. The Battle Bus appears to be a blue-painted, everyday school bus with a few important additions.

But what's really the deal with the Battle Bus? Is such a thing even possible?

What we know about the Battle Bus:

The body of the Battle Bus appears to be that of a fairly standard school bus. A quick online search shows that many people have already identified the style as that of a Blue Bird Bus—one of the most popular and common school buses in the United States.

So then, back to our question: could you really make a big old blue school bus, loaded down with 100 costumed warriors, actually fly? Well, seeing as how the Battle Bus appears to be just a hot-air balloon with an extra-large (and extra-heavy) "basket" (i.e., the bus), we'll want to look at what makes hot-air balloons fly in the first place.

HOT-AIR BALLOONS

The basic principle of a hot-air balloon is pretty simple: hot air is lighter than cold air, so when you fill a balloon with hot air, it rises.

In order to heat the air in a hot-air balloon, you need some kind of fire, which on most modern balloons is accomplished with the use of a propane burner positioned underneath an opening in the bottom of the balloon. When the burner is lit, it heats the air inside the balloon until it is light enough to lift the basket. The propane is usually stored in liquid form inside the basket. In order to keep it in a liquid form, this fuel has to be kept in a strong metal container under very high pressure. When the balloon pilot wants to lift off, they open a valve that causes the compressed liquid propane to expand into a gas and flow out of a burner nozzle, where it is ignited by a pilot light (which is basically just a small flame used to ignite gas).

On the Fortnite Battle Bus, you can see the burner quite clearly: it sits right smack in the middle on the top of the bus, adorned with lighted "V" logo (for Vindertech Industries, of course), and usually has a large blue flame coming out of the top of it.

The balloon part of an actual hot-air balloon is called the "envelope." When the first hot-air balloons were constructed by the Montgolfier brothers in France in the late 1700s, they made the envelope out of paper and a rough natural fabric called sackcloth. They also put a goat, a chicken, and a duck in their balloon before they would put any humans in it. Today, most hot-air balloons are made from nylon, and the vast majority of commercial balloons have a strict no-farm-animal policy.

Next we come to the basket, which is really where the Fortnite Battle Bus departs from the known universe of hot-air balloons. Generally speaking, balloon-makers try to make the basket as light as possible, so that the hot air will have less work to do to lift it off the ground. After all, the heavier the basket, the more hot air you need to lift it, and in order to get more hot air, you need to create bigger and bigger balloons and use more and more fuel to heat the air inside it. So, in the real world, balloon-makers generally use things the lightest materials they can find to make their baskets (older balloons mostly used wicker, but today plastics are more common). In any case, one material they don't often use is steel, which is exactly what your typical school bus is made of.

Now I know what you're thinking: *obviously*, they couldn't use a dinky little wicker basket, because they have to fit 100 battle-hardened Fortnite warriors (some of them in the shape of bananas or robotic lizards) into it.

This brings up another, slightly unrelated question: Can you even fit 100 battle-hardened Fortnite warriors into a standard, everyday school bus? My first thought was that this was more than a little implausible, but after a little research, I discovered that it is actually more than plausible; it is downright possible. Most school buses have twenty-four seats for passengers, which means that you can easily (and even comfortably) fit forty-eight people at two people per seat. While this is certainly enough for a little Team Rumble action, it doesn't quite cut it for a full 100-person Battle Royale.

Okay, so let's say we put three people in each seat. It would probably be a little tight (and from the looks of some of you people, it wouldn't really smell that great either), but the whole trip is over in about thirty seconds, so it's doable. Still, that only gets us to seventy-two people, leaving twenty-eight to stand in the aisle. Now the exact length of the aisle of a school bus can vary a little, but most of them tend to be around nine to ten meters long. If we assume the average Fortnite player standing will take up about a third of a meter (curved torsos like those belonging to Peely notwithstanding), standing single file in the aisle, pressed up against each other, twenty-eight people would need only 9.33 meters . . . clearly enough room for the short journey over the island.

In fact, according to the *Guinness Book of World Records*, the most people ever successfully crammed into an unmodified school bus was a whopping *229*. This was accomplished by the mechanical engineering faculty of the Krakow University of Technology in 2011. Amazingly, they actually managed to drive the bus a whole 75 meters in a little under a minute with all 229 passengers inside.

Okay, so now that we've got all 100 players on the bus, would it actually be able to fly? To figure that out, we need to consider a few things about lift. As discussed in the chapter on Skydiving and Gliding, lift is the force that pushes things like planes and hot-air balloons (and, in our case, buses) upward, allowing anything that flies to *lift* off the ground.

Lift is generally measured in pounds, which is quite convenient, because that's also how we measure how heavy people and buses are. So, in order to figure out if a Battle Bus could fly, we just need to figure out how many pounds a fully loaded Battle Bus would weigh, and then see how many pounds of lift a balloon can generate. If the lift is greater than the weight of the bus, then the bus will, in fact, fly.

First, then, we need to figure out how much the bus itself would weigh. In real life, you can find lots of different buses with a lot of different weights, depending on how many seats it has, what kind of engine is in it, as well as lots of other factors. With all these different factors, school buses will usually weigh somewhere between 23,000 and 30,000 pounds.

Now, on all school buses, one of the heaviest components will always be the engine, coming it at around 1,500 pounds. The exact weight of a school bus engine will depend on the model of bus and a few other factors, of course, but 1,500 is an acceptable estimate for our purposes here. The purpose of a school bus engine is to move the wheels of the bus 'round and 'round (as the song goes), which is hardly necessary when the bus is a few hundred meters in the air, so let's just go ahead and assume that Vindertech scrapped the entire engine before trying to get the bus off the ground. While we're at it, we might as well assume that pretty much any parts required solely for driving have been ditched (including brakes, shocks, transmission, etc.) and that, as a result, the geniuses at Vindertech have managed to get the total bus weight down to right around 20,000 pounds.

Of course, it's not just the bus that you need to worry about. You also have all 100 of those pesky Battle Royale warriors. Despite the age/weight/species of the person holding the game controller, it seems safe to say that pretty much all of the actual Fortnite warriors are adult-sized, humanish beings of one kind or another. So, if you assume that you've got about fifty male and fifty female warriors, including the non-humans (and that the lizards, bananas, or dopey-looking sasquatches are roughly the same size as their human equivalents), then we can assign an average weight of 165 pounds per warrior. This means we have to add another 16,500 pounds on the bus . . . and that's if they're all naked. Once you put on armor and clothes and cute pink dogs in backpacks (don't mess with my Dodger), you need to figure it all to be at least 17,000 pounds.

This brings us to a grand total of 37,000 pounds for the fully loaded "basket" of our hot-air balloon. Unfortunately, here's where the math gets a bit more complicated.

If you heat the air inside the balloon by 100°C then each cubic meter of air inside the balloon will weigh about 20 percent less than each cubic meter of the air outside the balloon. This means that you would need about 750 cubic meters of air to lift 1,000 pounds. Taking this up to the scale of our fully loaded "basket" means that you would need approximately 28,000 cubic meters of air inside the balloon to lift the Battle Bus.

To give you a sense of just how big that is, consider this: holding on to our basic assumptions about the Battle Bus itself, we would have to estimate that the volume of the existing balloon is around 250 cubic meters, meaning that the balloon would need to be more than 100 times larger than it currently is to actually get the bus off the ground. Clearly, the balloon we see hovering over the Battle Bus is nowhere near this big.

So, what are we left to assume? That the game is just lying to us? That the Battle Bus is purely imaginary? Well, not quite (at least, not quite yet). There are a few other numbers we can play with to try and make this crazy machine get off the ground.

The first thing we can do is to make the air inside the balloon even hotter, just because we can. Let's say that we heat the air inside the balloon all the way up to 225°C (because that's about the maximum temperature a standard nylon balloon can handle without melting). This means, if we apply the same math we did before, we would only need about 13,825 cubic meters of hot air to lift our fully loaded Battle Bus. While this is certainly better than 28,000 cubic meters, it would still require a balloon far bigger than the bus itself, and nothing like the relatively small balloon we clearly see in the actual game.

So, *now* is it time to give up? Well, yes and no. I mean, if we are assuming that a large banana can shoot an assault rifle, then we can also (I think) assume that Vindertech can invent some better ways to fly a Battle Bus than just sticking a regular old hot-air balloon rig on top of a regular old school bus.

Consider the limiting factors here:

1. **The material the balloon is made out of.** Nylon melts at about 225°C. But seriously, do you think Vindertech is going to use nylon? Of course not. Let's assume that they've created a material with a melting point of 1,000°C. This is quite possible, considering that NASA has been using stuff like that for years.
2. **The fuel for the fire.** Typically, hot-air balloons use propane for fuel, which burns at about 5,072°F. However, we already know that the Battle Bus doesn't use propane, because if you've played even

a few minutes of *Save the World*, you know that the Battle Bus actually runs on a mysterious substance called "BluGlo." Now, if we set aside the pesky fact that BluGlo doesn't actually exist, we can just assume that it's a super-awesome fuel source, and that it burns at about 50,000°F. Hotter fire means hotter air, and hotter air means more lift. Having more lift, in turn, means you can use a smaller balloon.

3. **Passengers.** As we all know, the players in this game do not arrive on or leave the island by means of a bus or a train or anything like that. In fact, they just seem to appear in a flash of blue light from out of that little flying robot thingy, which also takes them away when they die. This means one of two things: Either that little robot is capable of transferring matter into energy, transporting it wirelessly, and then re-forming it into matter (also known as teleportation), or else (just go with me here) maybe *the players themselves are really just holograms*. And if the players are holograms, then they are actually made completely out of light, and hence weigh nothing at all! Boom—we just took 17,000 pounds off our Battle Bus!

4. **Gravity.** Now, I don't know about you, but I can jump maybe a foot in the air on a good day. But in Fortnite, every player routinely jumps nearly five feet like it's nothing. Of course, this might just be the whole hologram thing, but then again, maybe it isn't. Maybe, in truth, there is actually *less gravity* on the Fortnite island than on the surface of whatever planet you live on. (Actually, if you read the chapter where we talk about gravity in Fortnite, you would remember that it's about 2.8 times greater than the gravity on Earth, so let's just forget we mentioned that one for now and stick with standard Earth gravity.)

Even if we manage to take out 40 percent of our starting weight, we can surmise that we would need to heat the air to more than 6,600°C to generate enough lift with such a small balloon. While I have a lot of faith in Vindertech, even this might be a bit too far outside of even their abilities.

So here's my hypothesis: Let's say that Vindertech did manage to create a material lightweight enough to make a balloon capable of withstanding temperatures up to 10,000°C, and let's also say they are burning pure BluGlo, which is easily able to heat the air in the balloon to 6,600°C. If they were willing to go that far, then we can assume that they'd be willing to build a custom bus out of something other than the steel and iron and other heavy metals that a regular school bus is made out of. After all, we don't build planes out of the same material we build school buses out of, do we? Of course not. School buses are made to be strong and sturdy and able to withstand car accidents without hurting the children inside, so they make them out of much heavier materials, most notably steel.

Vehicles that are made to fly, however, are always made out of the lightest materials that can offer enough strength to withstand the stresses of flight. For example, most airplanes today are made out of aluminum, which is three times lighter than steel. This means that if we assume an average school bus (stripped down, of course) weighs 20,000 pounds and is made up of 80 percent steel, and we replaced all of the steel with aluminum, we'd bring our total weight down to 6,000 pounds. Of course, there's no need to stop there. Real-life aircraft materials like graphite-epoxy can weigh as little as half as much as aluminum (and one-sixth the weight of steel). So, if we replaced all of the steel with graphite epoxy, we would take our total weight down to a paltry 5,000 pounds!

Alas, that still probably won't get us all the way to plausible, at least not with the tiny balloon we have come to know and love. Based on the calculations of one person on YouTube, the balloon on the actual Battle Bus is roughly 250 cubic meters, and with a standard hot-air setup, it would be capable of lifting 300 pounds at the most. If we assume that Vindertech can increase the efficiency of the balloon by 400 percent with fancy materials and a BluGlo burner, then we would need to have a bus that weighs less than 1,200 pounds for this whole thing to work.

And the only way to do that, I think, would be to basically just make the whole thing out of paper. So that's my answer, and you can consider it canon, factual, proof-positive.

The Fortnite Battle Bus is made out of paper.

RIFTS

Does anything like a Rift exist in the known universe?

If you are as lousy a Fortnite player as I am, then you probably think of Rifts (and especially the To-Go variety) as your favorite quick escape plan for when you are unfortunate enough to actually run into another player on the Island. But regardless of how you utilize the Rift, it is undeniably one of the most useful aspects of the game.

As you know, there are two types of Rifts in Fortnite: the environmental Rift, and the Rift-to-Go. The Rift-to-Go works basically like a portable, temporary environmental Rift, but both Rifts basically operate the same—when a player walks into one, they are instantly transported to a height of 250 meters over sea level, allowing them to skydive and/or glide to a new location on the map.

The Rift itself is a kind of glowing blue flower-shaped crack in space-time that makes a weird splitting sort of noise when you get close to it and a giant smacking sound when it's transporting you into the air. While the environmental Rifts remain stable and present at a fixed location throughout a match (though they do disappear after someone uses them), the Rifts-to-Go (which look like one of those old lightning globes from the nineties that they used to sell at the Sharper Image) open up when triggered by a player and then close roughly thirty seconds later.

It's important, right off the bat, to make it clear that the Rifts in Fortnite are not, as far as we can tell, *teleportation* technology. This is vital. There is a big difference between a Rift and, say, a man-made device of some kind that actually breaks down matter into energy, transports that energy (or transports information) across some distance, and then reassembles it on the other side. The closest thing to that are the little teleportation robots that drop you into the waiting area, and then suck you up when you die.

No, for Rifts to be some sort of man-made teleportation device there would have to be a machine of some kind on both sides of the Rift, attached to some kind of power source, with wires, gadgets, and computers controlling the whole operation. It appears quite clear that a Rift is, for lack of a better word, some kind of natural phenomenon—caught in the wild in some cases, and "bottled" in a to-go container in others.

So, assuming that's true, and that the Rifts are some kind of natural phenomenon, what exactly are they, and how might they work?

To explain that, we have to take about five steps back and look at the big picture of the universe, specifically what Albert Einstein discovered about the universe, and how matter, space, time, and energy all interact with each other. He called this discovery the Theory of Relativity.

The Theory of Relativity is actually two theories—the Theory of Special Relativity and the Theory of General Relativity. Einstein's Theory of Special Relativity came first, and essentially makes two claims:

1. Nothing can travel faster than the speed of light.
2. Time moves more slowly as you travel faster through space, in essence stopping completely when you reach the speed of light.

If you're like most people, that first claim seems pretty easy to accept—the universe has a speed limit, and that speed limit is 299,792 kilometers per second.

The second claim, however, can be a bit tricky to wrap your head around. Without getting too deep into the physics and math of the whole thing, just try to think about it this way: imagine you had two identical twins born on the same day. For the purpose of this example, let's say they're born on January 1, 2020. Now, as soon as these twins are born, you stick one of them into a spaceship that spends the first year of that child's life traveling in a big circle at 275,000 kilometers per second (this is a little more than 90 percent of the speed of light, and obviously quite impossible with current technology, but useful here for our purposes), while the other twin stays on Earth. Assuming the twin on the spaceship stays at exactly the same speed for the entire trip, and comes back to Earth exactly one year after they take off (according to their watches and calendars on the spaceship), the twin on the spaceship will be exactly one year old, right? Of course. Here is the interesting part, though: the twin who stayed on Earth will not be one; he will be nearly two and a half years old.

This is not science fiction (apart from the bit about the spaceship traveling at 90 percent the speed of light). This is verified, experimentally tested, and proven science. If it were not true, then the famous equation $E=mc^2$ would also not be true, and we wouldn't have been able to invent atomic bombs or nuclear energy.

But still, what does all this relativity stuff have to do with Fortnite Rifts? We're getting there.

The next part of Einstein's discovery came later on, and it's called General Relativity. This is where things really get interesting. When Einstein published his Theory of General Relativity, he was overturning centuries of scientific understanding about our universe, particularly as it relates to gravity. What Einstein managed to prove was that not only is time relative, time is actually its own dimension, just like the three spatial dimensions of length, width, and height. Combined, these four dimensions comprise what is known as space-time, and space-time is the real key to making the Fortnite Rifts work.

Remember way back in the chapter on gravity, when we made that huge rubber trampoline? No? Okay, well I really don't feel like writing all that again, so why don't you just go ahead and reread it. I'll just wait here.

All done? Good. So: you now recall how mass actually bends and distorts space-time, and that this distortion causes gravity, and how you can picture this as a bowling ball sitting on big rubber trampoline. Thus, when you try to roll another, smaller ball past the bowling ball, the smaller ball will get pulled toward the bowling ball by the curvature of the space-time. This is how gravity works. Objects with a large mass (like our sun) actually bend space-time, causing smaller objects to fall toward them as they travel across space-time. This is why the Earth is attracted to the sun, and why we don't just float right off of the surface of the Earth and drift out into space.

Now that you have that picture in your head, imagine that, instead of a bowling ball, you had a tiny marble-sized object, maybe one-half inch in diameter, but that weighed as much as one million bowling balls. What would happen if you dropped *that* on the rubber trampoline? Well, obviously, if whatever was holding up the sheet managed to stay standing, the marble would bust a hole right through the trampoline. Then, any other objects that rolled over that hole would fall in, disappearing right off the face of our space-time rubber sheet. You can think of this as sort of like a black hole—an object of such enormous mass that it bends space-time so much that literally everything that comes close to it—even light itself—falls in.

Now, let's take it one step farther. Let's say that right before you dropped this massively heavy, incredibly small marble onto the rubber sheet, you folded part of the sheet under the spot where the marble would drop. Now, instead of simply forming a hole and dropping into nothingness, the marble creates two holes—one going in and one coming out—in different places on the rubber sheet, which are actually different places in space-time. Then, if you were just a friendly everyday human-sized banana with an assault rifle, and you just happened to jump into one of those holes, you would instantly come out the other side.

In nature, this idea (which is, in truth, only a hypothesis, never having been experimentally proven) is called a wormhole. But it's basically the same thing as a Fortnite Rift—a hole in one part of space-time that leads to another part of space-time.

It has also been theorized that wormholes, should they actually exist, would in essence be a portal to a fifth dimension, which would go along quite well with the way that Rifts worked when they were first introduced in Season 5 (remember that whole alternate-dimension thing with the cube and everything? I'm still shaken up by that one) and swallowed up much more than individual players.

In fact, when Inverse.com journalists interviewed famed physicist Neil deGrasse Tyson about the idea, he said that the idea of Fortnite Rifts being wormholes was not that far-fetched at all. Actually, the thing he took the biggest issue with was the way the Rifts look. In reality, if such a wormhole were to exist, and you could use it the way that players use Rifts in Fortnite, then it would look more like a spherical hole, with the other side possibly even visible through the aperture (so then, closer to Portal than to Fortnite).

There are, of course, a few problems to consider when exploring the idea of Fortnite Rifts as wormholes. First and foremost, there is the obvious question—where do they come from? Are they naturally occurring on the island, or are they somehow manufactured by those evil geniuses at Vindertech?

This could really go either way. On the one hand, the technology necessary to manufacture wormhole rifts on demand is far beyond anything the human race can currently achieve. However, the same could be said for hoverboards and Boogie Bombs, so that's really neither here nor there. Still, I would be tempted to throw my hand in with the naturally occurring side of things if it weren't for two facts:

1. Some of the Rifts are contained in to-go containers.
2. We *might* already have the technology necessary to create a wormhole.

Here's how this all breaks down: If the Rifts were naturally (or at least randomly) occurring on this particular island (perhaps as a side effect of whatever the hell kind of experiments they're doing under the Dusty Divot), then they would probably only exist in the places where they form. Imagining a Rift forming naturally on an island is one thing, but imagining such Rifts forming, and then finding them and bottling up lots and lots of them, just doesn't seem very likely. On the other hand, if Vindertech created these Rifts inside some kind of controlled environment, they could conceivably figure out a way to halt their creation just a millisecond before its completion, and then trap it inside that little Sharper Image snow globe.

Sound far-fetched? Well, maybe it is, but not as much as you might think. A long time ago, back before there even was such a thing as Fortnite, a group of scientists in Switzerland created a machine called the Large Hadron Collider (LHC) that just might be capable of pulling off such a feat. The LHC is the biggest and most powerful particle accelerator in the world. A particle accelerator is basically a long (in the case of the LHC, nearly seventeen miles long), donut-shaped tunnel filled with powerful superconducting electromagnets. The purpose of these magnets is to guide a pair of particle beams in opposite directions around this donut-shaped tunnel, accelerating them faster and faster until they almost reach the speed of light, and then smash them together.

There are lots of interesting things you can learn from smashing particles together like this, but one of the most interesting discoveries the scientists made was that, mathematically speaking, there was a small chance that smashing certain particles together like this would actually create a wormhole right there in Switzerland. Of course, there was also a small chance that the LHC could create a black hole that would swallow up our entire solar system, but luckily that hasn't happened yet.

Still, it's not so hard to imagine that, with just a dab of fiction added to the science, those wily Vindertech physicists, working late into the night at the Vindertech particle accelerator, accidentally created a wormhole while trying to do something completely different (maybe trying to get a tomato to grow in the place of a human head, for example). Then, just maybe, that accidental wormhole opened up a "rift" in space-time between the center of the particle accelerator and a spot about a hundred meters in the air above them. One could even imagine them figuring out a way to freeze this reaction in the fraction of a second between the collision of these mysterious particles and the formation of the wormhole by encasing the whole thing inside some kind of special vacuum. So when the seal is broken on this spherical little vacuum, the reaction completes and the wormhole is formed. Though, of course, these portable wormholes would be far less stable than naturally occurring wormholes, so the Rift-to-Go wormhole collapses a few seconds after it forms.

Okay, yeah. That is a little too far-fetched. Oh well.

THE VOLCANO

From the moment it first exploded out of the ground in season, the Volcano became one of the most recognizable geographic features of the island. At first, it had a typical lava pit in the middle (also known as the caldera, but we will get to that later), which for some reason housed some sort of royal throne room with a few token loot chests scattered around it. Since then, the Volcano has played a significant role in the development of the island, from the rumbling and grumbling, to presumably being the gateway for covering the entire island with lava, to its current incarnation (at least as of this writing) as a "pressure plant" of some kind.

So, what is the science behind the Fortnite Volcano? How do volcanoes work in the first place, and is there anything about this particular volcano that goes beyond the volcanoes as we know them? Most importantly: what the heck is a pressure plant?

To begin answering these questions, we must first look at how volcanoes typically form on Earth, and to understand that, you first need to understand what's under the Earth's surface. If you were to slice the Earth exactly in half so that you could see all of its different layers, you would see that most of the inside of our planet is a glowing ball of magma, rock that has become so hot that it has actually melted.

One quick note: You may have heard the words "lava" and "magma" used interchangeably from time to time (though certainly not by me) and found yourself wondering if they really are the same thing. The true difference between the two is very complicated and quite subtle: both lava and magma are melted rock. While the exact kind of rock varies, it is usually basalt, and the actual composition isn't what differentiates the two. The real difference is this: when the melted rock is *inside* the Earth, it's called magma. When it comes up through the ground and makes it to the *surface* of the Earth, it's called lava. Got that? Good.

On top of this layer of molten rock are the plates of the Earth's crust, which float on top of the magma almost like a large sheet of ice might float on top of the ocean near the North Pole. Because they are floating on a sea of liquid rock, the crust plates are constantly in motion. It doesn't feel like it to us because the motion is very slow—the tectonic plate on which the United States sits moves approximately one inch per year, far too little movement for you to notice while you are just standing in your backyard.

Sometimes, however, you can feel the movement of these large plates of rock as they float on a sea of liquid magma. Each plate moves in a rather halting and jerky fashion. If you want to feel what I mean, try pressing your hands together as hard as you can, and then moving them in opposite directions just a little bit without taking any of the pressure off. You will notice that at first your hands do not move at all, but when you finally build up enough pressure, the friction breaks and they move very quickly a short distance before stopping again. This is how tectonic plates usually move, and when the edges of two plates rub up against each other and vibrate everything around them, it's called an earthquake.

Sometimes, where these plates overlap and are brought up against each other, small openings form leading down to the sea of magma below the plates. This magma is under so much pressure that when such an opening occurs, the magma starts to rise out of it, eventually making it all the way to the surface. This is a volcano.

There are lots of different types of volcanoes. Some, like the ones that make up the archipelago of Hawaii, dribble lava very slowly. This happens when there is less pressure, and there are enough openings so that the pressure of the magma is released in small bits over time.

Other times you will have a volcano that builds up pressure over long periods of time until it becomes too great for the land above it to maintain. An explosion takes place, usually blowing the entire top of the volcano and forming a giant crater, like the one you currently see on the Fortnite island.

Speaking of violence, it is entirely likely that the Fortnite island itself was created by the volcano that sits on top of it right now. This is an obvious conclusion, because many islands throughout the world were formed by volcanoes: at the bottom of the ocean, a small opening forms between the seabed and the molten rock layer beneath the Earth's surface. Out of this hole pours, at first, just a small amount of lava. Now, being liquid rock, when the lava touches the cold ocean water, it quickly solidifies into hard rock. Then a little more lava comes out and piles on top of that rock before solidifying, making a larger and larger pile of cooled rock on the bottom of the ocean floor. As the lava continues to come out of the top of this pile and flow down the sides, an underwater mountain is formed. This mountain grows bigger and bigger over the years until eventually it reaches the surface of the ocean. At first it is just a tiny little island with a tiny little volcano right in the middle. But as time goes on and more lava comes out the top of the volcano, the island will grow. The wind and rain and ocean will bring different types of plant and animal life to the now-exposed surface of this mountain, and over time it will begin to look much like what we think of as a volcanic island today.

Now that you know some of the basics about what volcanoes are and how they are formed, what about the Fortnite Volcano? Is it more or less realistic, or is there anything about it that couldn't really exist in the real world?

The first thing that definitely doesn't translate very well into the real world is the lava itself. In Fortnite, when a person (or banana, or dog-man, or whatever) jumps on a river of lava, they merely bounce around, while presumably getting very mild burns on their feet (I assume this is true based on the relatively small amount of damage you receive when you jump on the lava). Is this realistic? What would actually happen if you were to jump on lava? Would you really bounce like a rubber ball, or would you just, you know, die?

Believe it or not, this one is actually pretty easy to answer. When lava is first expelled from underneath the surface of the Earth, it is heated to a blistering temperature of approximately 700–1,200°C. Now, if you were to jump on a river of lava (which I really suggest you avoid doing, if you can), there would be two or three layers of material between your skin and the molten rock. Of course, this all depends on what you are wearing that particular day, but let's just go ahead and assume you have on a pair of sneakers and some basic cotton socks.

The first thing that's going to come in contact with the lava is the soles of your shoes, which we're going to assume are made out of pure rubber, which has an ignition temperature of approximately 260–316°C. Match that up against our 700-degree (minimum) lava, and your shoes don't stand a chance.

Next, depending on your shoes, you may have some leather, or else just your cotton socks. Not that this really matters: the ignition temperature of both rubber and cotton is less than 250°C, so that's all pretty much toast, as well.

Finally, all you are left with is your skin, and seeing as that skin starts to burn at far lower temperatures than rubber, leather, or cotton (about 44°C), there's really not much hope for the jumping-on-lava thing being true.

Of course, it's possible that all of the players in Fortnite are given lava-proof shoes, but even if they are, this would not be something that would translate to real life at all. Of the very few materials on the Earth that can withstand the temperature of flowing lava, absolutely none of them make particularly good material for shoes.

Second thing: is it possible to generate power or electricity with a volcano?

What is that pressure plant in the Volcano anyway? There's not a whole lot that can be told simply by looking at it, though the clear indication is that it is some type of power plant running off of the Volcano's energy. Of course, this being Fortnite, it could turn out in a few seasons that the pressure plant is actually, I don't know, a giant alien tomato castle (or whatever), but for now let's assume that it is at least *supposed* to be a power plant at some point. So, can you build a power plant out of a volcano?

To answer that question, we will first begin by looking at geothermal power plants. These are in fact very real, and an excellent source of sustainable energy. But what is a geothermal power plant, exactly? The answer to this question is in the name: *geo* means Earth, while *thermal* means heat. So *geothermal* means heat from the earth, and a geothermal power plant is a facility that harnesses heat from the earth in order to make electricity.

This type of power plant actually works quite similarly to other types of power plants. Most power plants, the kinds that burn fossil fuels, work by burning something like coal, in order to heat up water until it is boiling, and then placing a fan over the boiling water so that the steam turns the turbine, which turns the copper wires, which makes electricity (see the chapter on Driftboards for an explanation of how power plants generate electricity with copper wire and magnets).

Normally, to make this happen, a power plant will burn coal, or perhaps natural gas, or even create a nuclear reaction to make the above process happen. With a geothermal power plant, the process is essentially the same; however, both the heat source and water source are already waiting to meet the Earth.

As we discussed, sometimes magma under the earth creates caverns and pockets where it builds up pressure. Sometimes these caverns and pockets will come very close to a water source of some kind that is also underground. When this happens, the magma will heat the water to its boiling point and create a large volume of steam. With a geothermal power plant, there will be a drill of some kind that will puncture the cavern where all of this steam is being created by the magma and the water, and harness it into a turbine that turns a generator. These kind of power plants, however, are not generally made on active volcanoes, and certainly not inside the caldera of an active volcano. But why not? And could any kind of power plant ever be made someplace like this?

While the basic principles that make a geothermal power plant function could certainly work in such a situation, there are some mitigating factors that make such an application unlikely. First of all, there is the general danger of the whole thing. We still have no truly accurate way of predicting when an active volcano will erupt, so anyone building or working at such a plant would quite literally be risking their life every time they went to work. Of course, they probably wouldn't ever get that far, due to one very serious snag in this whole plan: every geothermal plant needs to have a few holes dug down into the earth to move water down toward the heat and then to bring the heated steam to the surface. The only way to dig holes like that is with a drill. However, there isn't a drill on Earth (at least that anyone has managed to invent yet) that can drill into molten lava. Any drill they tried to use would just melt. Finally, of course, is the practical answer to this question, which is that there just wouldn't ever be a reason to build a power plant right in the caldera of an active volcano. Pretty much any volcano like that is going to be surrounded by plenty of more deeply buried, not-about-to-blow-up magma pockets that would make for a much safer and more feasible geothermal power plant.

If I were to guess, I'd say you'd find a great spot right between Loot Lake and the Volcano where you could get your water from the lake and then dig down deep enough to heat it up with a nice, stable magma pocket.

THE STORM

So there you are, happily searching some secret little spot on the island for some sweet loot at the beginning of the match. Maybe you've found a few choice weapons, maybe you're just taking in the view from the top of the Volcano, or maybe you're just trying out some new dance moves in the Dusty Divot. Whatever you're doing, though, you are not paying enough attention to the sky, because all of a sudden you catch a glimpse of the mini-map in the corner of the screen and you see it: that big purple blob of death, slowly lumbering across the landscape toward you.

That's right: it's the Storm, towering from the ocean all the way up as high as you can see in the sky, that big swirling purple monster that strikes fear into the heart of every Fortnite player.

So what do you do? Well, if you're anything like me, maybe you haven't even checked to see where the safe zone is yet. So you look around to see if there is any readily available mode of transportation to jump on: a Quadcrasher, a Driftboard, anything . . . But there's nothing. You are alone, and the Storm is coming for you.

So you start to run. Blindly, you follow the mini-map's little white line that is supposed to lead you to safety. At first, it seems like you just might be able to run fast enough to beat the leading edge of the Storm, but you soon realize that any such hope is in vain. Maybe you make it over a few hills, maybe you can even see the safe zone off in the distance, but before you can get there, it's already too late. The Storm has overtaken you. All of a sudden you are surrounded with loud, crackling, purple death.

Your shields won't help you, so you try grabbing an apple off of the ground, but it barely makes a dent in the health you're rapidly losing, if you don't hurry up and get out of here, your whole match will be over before it even starts. Wildly, you run toward the safe zone as you watch your health slowly draining away. Finally, you see the barrier of the Storm ahead you. The eye of the Storm is within sight. If you can only reach it, you are sure you can find some way to heal yourself on the other side: a med kit, some bandages, some apples, or a fire. But just when you are a few steps away from that glorious line of safety, the Storm takes its last few stabs at your health, and it's all over.

Anyone who has played any amount of Fortnite will probably be able to relate to this scenario. The Storm is an ever-present, extremely important part of what makes Fortnite such a great game. Shortly after the beginning of a match, a safe zone is defined by a circle on the map, and the countdown begins before the Storm starts closing in. It is essential to the game, as it moves every player toward a final endgame position, where no matter how well they may hide, they will be thrown right into the same small circle as everyone else who is still alive. It is not an exaggeration to say that you really could not have a *Battle Royale*–style game without a Storm (or at least something like a Storm) to push the action toward its ultimate conclusion. But what is the science behind the Storm? Is this kind of deadly storm possible in the real world, or is it just one more wild, weird, impossible part of the Fortnite universe?

To answer that question, we're going to need to look at two different issues: how large storms like the one in Fortnite come into existence in the real world, and what could make a storm (or any kind of precipitation) more dangerous than simply water falling out of the sky mixed with a little bit of wind.

First, let's talk about how storms are formed in the first place. The Storm in Fortnite looks very different from any kind of storm you're ever going to see in the real world, obviously. But if you're going to find anything even a little bit similar, it would have to be a hurricane.

A hurricane is another name for the type of storm known as a tropical cyclone. The basic way that these storms form is actually quite simple. Near the equator, where the ocean is the hottest, the air on top of the ocean becomes warm and moist, and like anything that is warm, it starts to rise. As this warm, moist air rises higher into the atmosphere, it leaves less air down near the surface where the ocean is. When there is less air in a particular area on the surface of the Earth, that is known as a low-pressure area: literally, there is lower air pressure there than in the areas surrounding it. Anytime you have low air pressure next to places with higher air pressure, the air from the higher-pressure locations starts to rush toward the lower-pressure locations to achieve equilibrium. This is what starts the process of wind blowing that a hurricane is known for.

When the air from the higher-pressure locations has made it into the area of low pressure, however, it warms up and gains moisture just like the air it was rushing in to replace. This starts the entire process of the hurricane. While the wind is starting to blow in and fill the low-pressure area, the warm, moist air that has been rising begins to cool as it reaches the upper levels of the atmosphere. When this moist air cools it forms into clouds, and when these clouds get dense enough, they begin to drop rain.

As this whole process picks up speed and energy from the huge supply of warm, moist water near the equator on the surface of the ocean, it begins to spin as a vortex is created. This spinning vortex of warm moist air is what creates the hurricane.

When there is enough warm air to act as fuel for this system, it will just keep on growing until it forms the spiraling hurricane that we all know and fear. Just like in Fortnite, hurricanes always have an eye: a circular area right in the middle, where there is no storm. However, unlike in Fortnite, real hurricane eyes do not shrink tighter and tighter while the storm keeps its overall strength. Generally, when a real hurricane hits land, and the eye is no longer on top of the ocean where it can suck up all of its warm moist air for fuel, the spinning center of the hurricane begins to fall apart, and the hurricane becomes eventually just another rainstorm.

But what about the real danger of the Storm in Fortnite? Obviously, a hurricane is dangerous because the wind is so strong and it often brings a surge of ocean water that floods the coast. In Fortnite, however, it is seemingly the rain (or maybe even the air) that is so dangerous. Is there anything like that in real life?

Probably the closest thing you're going to find to this type of rain in the real world is acid rain. So what is acid rain?

First of all, real acid rain does not necessarily have to be rain at all. It can be snow, fog, or sometimes even just dry acidic material in the air.

But whatever form it takes, acid rain is any kind of precipitation that has extremely high levels of nitric and sulfuric acid.

For the most part, acid rain is caused by the burning of fossil fuels, coal in particular, for the production of electricity in power plants. These plants burn fossil fuels to turn their big turbines and create electricity, but the exhaust and gases that burning fossil fuels release contains quite a lot of sulfur dioxide and nitrogen oxides. These pollutants rise up from the smokestacks high into the sky, where they are dispersed in the upper atmosphere and mix in with clouds and then contaminate the rain that is falling out of those clouds.

Can acid rain really kill you the way that the Storm in Fortnite does? Well, no. Thankfully, acid rain is not nearly as dangerous as the rain that falls in the Storm in Fortnite. If it were, we would all be dead already. That does not mean that it is not harmful, however. Acid rain can be extremely detrimental to lakes and rivers and all of the wildlife that live there, and when those waterways become polluted enough, it can have an extremely detrimental effect on the entire ecosystem. But still, most people on Earth have had acid rain fall on them at some point or another, and have lived to tell the tale.

One final note about the Storm: while acid rain is generally created by pollution, it can also be caused by volcanic activity, as volcanoes do release many of the same noxious chemicals into the atmosphere that power plants do. So even though we can't find a scientific explanation for the purple swirling death storm that is the Storm in Fortnite, we can perhaps find a reason on the island for that acid rain to be there in the first place. And who knows, maybe some of the crazy things that the Vindertech folks are studying underneath the Dusty Divot have something to do with the especially nasty volcanic activity taking place near the pressure plant, and just maybe that is actually the cause of the super deadly acid rainstorm that we have come to know and fear.

PART 2
Shooting Science

HOW GUNS WORK

Nearly 1,200 years ago, ninth-century Chinese alchemists were searching for an elixir of life—a potion that would allow whomever drank it to live far longer than the medical science of the time could accomplish—when they happened upon a mixture of saltpeter (potassium nitrate), charcoal, and sulfur. To say that these particular alchemists failed at their appointed task would be the understatement of the millennium. Not only did this "potion" completely fail to extend the life of anyone, this particularly potent mixture of chemicals became responsible for shortening the lives of perhaps more people than any other invention in the history of humanity.

This invention was, of course, gunpowder: the combustible substance responsible for every firearm that has ever existed in the real world and, by extension, every firearm in Fortnite.

Before we start looking at the science behind all the weird and wacky weapons in Fortnite, we need to explore a little bit about how gunpowder works. What is it, specifically, about this happenstance mixture of three common, everyday chemicals that led to the creation of everything from a pirate cannon to the SCAR assault rifle?

As mentioned above, basic gunpowder is a mixture of potassium nitrate, charcoal, and sulfur. Contrary to popular belief, gunpowder does not really explode, exactly (at least, not in the same way that something like TNT explodes). Rather, the combination of these three chemicals, once ignited, simply burns at an extremely high rate of speed. Under certain conditions, this fast-burning powder can *seem* to explode, but we'll get to that shortly. First, we need to look at the three chemicals that make up gunpowder.

Let's start with sulfur. Sulfur is the same chemical that you find on the tip of a common matchstick. If you've ever seen anybody light a match, you know that once it sparks just a little, the rest of the sulfur burns very quickly.

Charcoal is also something you have seen often, most likely in your backyard barbecue. Charcoal is a very dense carbon substance, but provides a lot of potential energy for a fire.

So, with sulfur and charcoal, you have two different flammable substances mixed together—one that burns very quickly, but doesn't contain a great deal of energy, and one that burns more slowly, but does contain a whole lot of energy. What happens when you mix them together? Clearly, you will get a substance that burns quickly *and* has a lot of energy, right? Isn't that basically what gunpowder is? And if so, what do you need with a third ingredient?

Sure, if you mix sulfur and charcoal together in a big pile and then toss a spark on top, you're going to get a nice big fire. But you won't get *nearly* as big a reaction as you will once you add potassium nitrate.

As we discussed in the chapter on jetpacks, any kind of fire needs both fuel and oxygen to burn. So if you were to mix sulfur and oxygen together and set them on fire, your little inferno would need to pull in oxygen from the environment surrounding it to achieve ignition. This will certainly work if all you want is a fire, but the process of the fire pulling in all that oxygen from the air is actually quite slow. That's where the potassium nitrate comes in.

Potassium nitrate is an oxidizer, which essentially means that it acts as a source of oxygen for a chemical reaction. In the case of gunpowder, potassium nitrate is an especially effective oxidizer, because, as a crystalline powder, it provides the oxygen needed for combustion without the need to pull in oxygen from the atmosphere.

So, by mixing a substance with a lot of stored potential energy (charcoal) with a substance that burns very rapidly (sulphur) and a source of easily available oxygen (potassium nitrate, a.k.a. saltpeter), you wind up with a powder that burns faster than nearly anything else invented by man. If you were to put a pile of gunpowder on the ground and light it with a match (please do not do this), it would not, however, create an explosion. Why not?

Due to the rapidity with which it burns, gunpowder actually derives its power from the amount of gas and smoke it creates when it burns.

Just like anything else, when you burn gunpowder, you are transforming it from a solid into a gas, and like all gases, the gases from gunpowder take up much more space than gunpowder does in its solid form. In other words, igniting gunpowder makes its volume increase dramatically in a very short period of time.

Of course, burning anything makes its volume increase, as well. Let's say you have an ordinary wooden log that is exactly one cubic foot in volume. When you put the log in your fireplace and set it on fire, that log is transformed primarily into heat and gases (smoke, carbon monoxide, etc.) with a small amount of solid left over (ash). Pretty much every substance in the universe can exist at any given time (based on number of physical conditions) in one of three states: solid, liquid, or gas (yes, there is also plasma, but that's just going to confuse the issue right now, so let's not worry about it in this chapter). Water, for example, can either be ice, liquid water, or steam, depending on its temperature. If you put a piece of ice in a pan and put the pan on the stove, your ice will first melt into a liquid, and eventually boil off in the form of steam. The important thing to understand here is that regardless of what the material is, transforming it from a solid into a gas makes it take up a lot more physical space. With substances like wood and water, this is a relatively slow process. With gunpowder, however, this process happens almost instantaneously.

That said, even with gunpowder, if this transformation from solid to gas happens in a pile on the floor in your kitchen, you'll simply wind up with a large plume of smoke filling the room. But what happens if you contain the solid material in something solid, not flammable, and too small to contain all of the gases before you ignite it?

To answer that question, let me ask you another one. What happens when you shake up a can of soda? Well, when you do that, you are transforming some of the liquid carbon dioxide inside the beverage into gaseous carbon dioxide. And just like with anything else that you transform form a liquid or solid into a gas, you wind up making a smaller volume of liquid into a larger volume of gas. As you well know, this creates a great deal of *pressure* inside the can: sometimes so much that if you were to open the can under all this pressure, the contents inside would explode out forcefully into your face. Don't believe me? Go ahead and try it; I'll wait here.

Convinced? Good, because the same basic process that happens inside that soda can is what happens with gunpowder inside an enclosed container: the gas contents expand greatly as they transform into a gas, building up pressure inside the enclosed space until something allows that pressure to be released, often with a rather violent explosion of force.

CANNONS

To illustrate how the violent expansion of gunpowder gases works, we don't need to look any further than one of the oldest gunpowder weapons ever created (which also happens to be a rather unique and fun weapon in Fortnite). I am, of course, referring to the cannon.

Although it was not the first form of gunpowder weapon (that was something called a fire lance, invented in ancient China to fight off the Mongol hordes, and if anyone from Epic is reading this, it would really make an excellent weapon for Fortnite), the cannon is certainly one of the oldest and most straightforward. As such, it is an excellent place to begin our discussion of how guns work.

A cannon is essentially a long metal tube that is closed on one end and open on the other. At the back of the cannon there is a small hole, into which a flammable piece of rope or cloth of some kind is inserted, which comes into contact with a quantity of gunpowder that has been packed tightly into the back of the tube. In front of the gunpowder is the cannonball, which is sized to fit snugly (though not tight enough to stick) inside the tube, nearly (but not quite) creating a seal for the gunpowder in the back.

When the fuse is lit, the fire travels down its length, through the small hole at the back of the cannon, and then ignites the gunpowder. The gunpowder then burns so rapidly that it is almost instantly transformed from a room-temperature solid into a very large volume of hot gases. Gunpowder's temperature will increase to more than 3,000°C while the mixture of potassium nitrate, charcoal, and sulfur transforms into three different solids (potassium carbonate, potassium sulphate, and potassium sulfide) and three different gases (carbon dioxide, carbon monoxide, and nitrogen). Now here's the important part: the volume of these newly formed solids and gases will be over 3,600 times greater than the volume of gunpowder.

Of course, when this transformation happens, the gunpowder is wedged down between the back of the cannon's barrel and the cannonball itself. This creates an enormous amount of pressure behind the cannonball, much like the pressure inside a shaken-up soda can (though obviously far greater). In a matter of milliseconds, the pressure behind the cannonball gets high enough that is actually pushes the cannonball out of the open end of the tube at an extremely high rate of speed.

To figure out what happens next, you just need to remember Newton's second law of motion, which states that anything in motion will stay in motion unless acted upon by another force. If there is nothing in front of the cannon when it fires, the cannonball will stay in motion until gravity and the friction of air resistance eventually slow it down and it falls to the ground. However, gravity and friction are not the only forces that can slow down a cannonball. The entire point of a cannon is for the cannonball to hit something hard enough that the friction of the collision actually stops the cannonball, destroying whatever it collided with in the process.

This is the basic idea of how cannons (and, by extension, pretty much all firearms) work: gunpowder is ignited in an enclosed space, transforming it from three room-temperature chemicals with a small volume to six very hot chemicals with a very large volume, thus creating pressure that pushes an object out of the open barrel fast enough to damage anything it comes in contact with.

Reality check: can you really shoot a person out of a cannon?

Ever since the cannons were first introduced to Fortnite in Season 8, they have been just as useful (if not more useful) as a means for getting around the map as they have as an actual weapon (they're not that easy to aim, and accuracy is much less important when you're shooting yourself across a river than when you're trying to score an elimination from a hundred meters away). But is there any truth behind this? Is it really possible to launch a person out of a cannon?

Well, this question probably needs a little bit of clarification. In Fortnite, the cannons you use to shoot yourself across the map are the same cannons you use to shoot an actual cannonball. When you shoot a cannonball, the logical assumption is that the cannon works very much in the same way that all cannons do: gunpowder, ignition, expanding gases, etc. Furthermore, as far as anyone can tell from simply observing this feature of the game, there is no significant difference between how the cannon shoots the cannonball and how the cannon shoots a person. So the real question needs to be: can you shoot a person out of a standard cannon, like the ones in the Pirate Hideouts in Fortnite, using ignited gunpowder? Is this really how human cannonballs work?

No. First of all, there's the fact that the cannons in Fortnite are quite clearly nowhere near big enough to fit a human being in their barrels. Still, even if you set this fact aside, it is safe to assume that were you to stick an actual person into a cannon like this, pack them up against a large enough quantity of gunpowder to launch them hundreds of meters into the air, and then light it, your human cannonball would be more or less instantly killed. Typical gunpowder burns at more than 2,000°C. Meanwhile, human skin begins to burn at just 44°C, and water (which makes up about 80 percent of the human body) boils at 100°C. This means that even if your human cannonball withstood the shock of the igniting gunpowder (they wouldn't), they would be little more than a ball of charred goo by the time they landed on the ground.

Okay, so I know what you're thinking: you've probably seen people being launched out of cannons before (at the circus or on TV or in a movie), and they don't wind up as puddles of charred goo. If these people can do it, why can't we do it with the Fortnite cannon?

Human cannonballs do not use actual gunpowder to fire the person in the air. While each circus performer who does the human cannonball trick has their own unique method for launching, they never use real gunpowder. Usually, they rely on a sled of some kind, which is mounted on a track inside the cannon and attached to series of bungee cords. These cords are stretched tight before the trick and then released when the cannon "fires," slinging the sled toward the opening of the barrel with the performer simply holding on. When the sled reaches the opening of the barrel it stops abruptly, but the performer keeps going, successfully launched into the air. Often there will be a firework set off at the exact moment sled is released to give the appearance that gunpowder is being used, but this is just for show and does not affect the actual cannon at all. There are other methods that rely on things like compressed air or metallic springs, but the basic idea is the same.

All that said, even with a safe, non-explosive means for launching yourself out of a cannon, the real, "tricky" part of the human cannonball trick has nothing to do with getting yourself airborne, but rather figuring out how to get yourself back on the ground without, you know, dying. For this reason, performers who do the cannonball trick will generally make sure to land in a large net of some kind, and not slow themselves down by crashing through the wall of a building.

FLINT-KNOCK GUN

While the cannon was first used in warfare sometime around the twelfth century, it would take almost 300 years before we would start to see guns that look more like the handheld firearms that are common today and are most closely related to the majority of guns in Fortnite. One of the first of these modern firearms was called a matchlock gun, which had a long tube (the barrel) into which you would first place a small quantity of gunpowder, and then a metal ball in front of it. A match (which looks a little different than today's matches, but the idea is the same) would be inserted into the back of the barrel to ignite the gunpowder and shoot the projectile. This type of gun was not very effective or reliable, because not only do lit matches often go out, but getting it into the right position could also be quite difficult. For this reason, the design of the matchlock gun was improved into a new design, whose name might sound a bit familiar to you: the flintlock gun.

Yes, you read that right, the gun was originally called a flint*lock* gun, not a flint-*knock* gun (the "knock" being added by Epic, presumably because it knocks you back so far). The flintlock gun first appeared in the 1600s, and made one very important improvement over the matchlock gun. Instead of having to insert a lighted match into a hole that would ignite the powder, a flint was attached to a swinging arm on a spring over a flat piece of steel. So when the trigger was pulled, the arm would swing the flint against the piece of steel, which in turn created a spark that was directed right into the back of the barrel where the gunpowder sat. This spark ignited the gunpowder, which expanded into a gas, and shot the projectile out of the barrel. This gun was far more reliable than the matchlock gun, as striking flint and steel produces a spark even in windy conditions, and certainly with much more ease and consistency than lighting a match.

The design of the Flint-Knock Pistol in Fortnite is actually strikingly similar to a real flintlock gun called a flintlock blunderbuss (which is, in itself, a pretty awesome name for a gun), a rather rare type of pistol that was popularized by the British navy in the late eighteenth and early nineteenth centuries.

THE REVOLVER

While the flintlock guns were quite revolutionary in their time (historical pun intended, of course) and were greatly improved upon throughout their lifespan, they still had a number of significant drawbacks. They were slow to load, they would stop working if the hole the spark needed to go through became clogged with dust or dirt, and they were very susceptible to water damage. If even a little bit of water wound up touching the gunpowder, you wouldn't be able to fire the gun at all, which made using this type of firearm in the rain quite difficult.

This brings us to the next Fortnite weapon that we're going to look at to explain the science of firearms. I'm talking about (at least in my humble opinion) the absolute worst gun in all of Fortnite: the revolver.

The primary scientific advancement that made the revolver possible was the invention of the waterproof cartridge, which is essentially what you think of today when you think of a bullet. What made this invention so special was that it managed to get both the projectile and the gunpowder into a single waterproof container (the cartridge) that could be fired simply by striking it with a small metal hammer at a specific spot on the back.

So, how did this new waterproof bullet cartridge work? A modern bullet cartridge has four primary parts: the bullet, the casing, the gunpowder, and the primer. The primer is a small piece of metal and a combustible chemical at the very back of the cartridge. When it's struck with the gun's hammer, the primer sends a tiny spark into the cartridge, where the gunpowder has been tightly packed behind a metal projectile (the bullet). The bullet is sort of wedged in front of the gunpowder, held in place by the tension of squeezing it inside a precisely sized cartridge.

When the primer is struck by the hammer (also known as the firing pin), it ignites the gunpowder, which creates that huge expansion of hot gases right there inside the chamber and pushes the bullet out of the casing and down the barrel of the gun. This allows the revolver to fire without needing to create a spark in the open air, or to have gunpowder exposed to the elements. As such, it is much more reliable and far faster to load than the flintlock and matchlock guns that came before it.

The revolver also has another feature that made it much more efficient than all of the other guns that came before it, which gives the gun its name: the revolver has a rotating cylinder that holds six bullet cartridges and revolves to the next cartridge after each shot. This was a huge advancement, as it meant that someone could fire six times before needing to reload the gun.

Clearly, this is a huge advantage over a flintlock gun (which, as in Fortnite, needs to be reloaded after each time it is fired), and if Epic ever decides to make a limited-time mode in which there are only Flint-Knock Pistols and revolvers, the revolver might actually be worth using.

Alas, the great advantage the revolver had when it was first invented became its primary drawback when compared to more modern weapons (especially those you find in Fortnite): the speed of loading, and the number of rounds it holds. Once you go through all six rounds in a revolver, the empty shell casings have to be manually ejected from the gun, and each individual bullet has to be carefully placed in one of the six slots in the revolving cylinder. While this was much quicker than other nineteenth-century firearms, the time it takes to do this makes it a much slower gun to reload, and one that needs more frequent reloading, than most of the guns you will find in use today.

DPS

DPS, that oh-so-important number you see highlighted next to every weapon in Fortnite, stands for Damage Per Second. This statistic is arrived at by combining the amount of damage that each bullet inflicts on the player being shot with how fast the gun is capable of firing those bullets. So let's say we have two friends playing Fortnite (we'll call them Bob and Fred), and each one has a different gun: Bob has a hunting rifle and Fred has a submachine gun.

For the sake of easy calculation, we'll say that each bullet the hunting rifle fires inflicts 100 points of damage, and the rifle can fire one bullet per second. The submachine gun, meanwhile, only inflicts 25 points of damage with each bullet, but it can fire five bullets every second. Now, if Bob and Fred were to stand directly across from each other, just a few feet apart, and begin firing at the exact same time, landing body shots with each and every bullet that they fire, who would win?

To figure out the answer to this question, we just need to calculate the DPS of each weapon. We do this simply by multiplying the damage by the number of rounds the gun can fire in any given second. The hunting rifle, firing one bullet per second at 100 points of damage, has a DPS of 100. For the submachine gun, on the other hand, we multiply 25 by 5, which gives us a DPS of 125. Thus, in this specific scenario, it is Fred, with his submachine gun, who will win.

What is the science behind the DPS for guns firing in the real world? Can you give an actual gun a DPS rating?

Well, yes and no. The damage that a real bullet will inflict on an actual living being is much more variable than anything that could ever be quantified with such a simple number. That said, we can examine the different factors that play into how much damage a given bullet can cause, and compare different bullets to each other.

There are two things we need to look at when considering this question: how much energy, or force, the bullet has as it's traveling through the air, and how much of that energy is then transferred into the target upon impact. In order to figure out how much energy a bullet has, you need to look at the caliber of the bullet, and the speed at which the bullet is traveling through the air. Caliber is generally a measure of the diameter of the bullet itself, and is typically a standard measurement. So, a .22 caliber bullet is .22 inches in diameter. This bullet will travel at different speeds depending on what kind of gun it is fired from. For example, if it is fired from a small handgun, its cartridge will contain a smaller amount of gunpowder, and therefore the bullet will travel slower than a .22 caliber bullet that is fired from a rifle, whose cartridge will usually contain a larger quantity of gunpowder. Obviously, the more gunpowder there is in the cartridge, the greater the volume of gas that is created when the gunpowder ignites, which will, in turn, propel the bullet forward at a faster rate of speed.

So for this question, we mainly want to look at the size of the bullet and how fast it is traveling. As both the size and the speed of the bullet increase, the damage the bullet can cause increases as well.

That said, this is only half of the issue. The damage a bullet causes is dependent not just on how much energy the bullet is carrying with it but, more importantly, on how efficiently the bullet transfers this energy into its target. For example, if a bullet is traveling so fast that it goes right through its target and then hits a wall behind it, a large portion of the energy that bullet was carrying is actually transferred into the wall, causing damage to the wall and not the target. If, on the other hand, a bullet with the same amount of energy manages to come to a complete stop inside the target, it will transfer all of its energy into that target and cause more damage as a result.

So how do you get a bullet to stop inside a target? This has to do with the way the bullet is constructed. A bullet with a sharper point, for example, is more likely to puncture through something than a bullet with a duller point. However, the sharper bullet will be more aerodynamic, allowing it to travel faster and therefore have more energy than a duller bullet. So, in order to get an aerodynamic bullet to stop inside a target, that bullet needs to expand or break apart upon its first impact. The bullets that cause the most damage are called hollow-point bullets, because the tip is actually hollow, causing it to flatten into a sort of mushroom shape when it hits a soft target. This flattening slows down the bullet and makes sure that it remains lodged in the first thing it hits, thus transferring all of the energy from the bullet into its target, causing the greatest amount of damage.

RATE OF FIRE

Why do some guns fire faster than others?

This is certainly one of the most important questions necessary to understanding the way modern firearms work, and the science behind them. Much of the improvement in firearms over the last 300 years has focused on the ability to fire rounds in faster succession.

As mentioned earlier, early matchlock and flintlock guns were very slow indeed. Because they required reloading after each shot, and reloading was itself a multistep process, even an experienced marksman could not shoot more than a few rounds per minute.

The invention of bullet cartridges, which fire when a spring-loaded hinge strikes a primer on the back of the bullet, took away the need to individually load each bullet from the front of the rifle and greatly sped up the process of firing. One of the earliest mechanisms for using this technology to increase the firing speed, which can be seen with the Hunting Rifle and some of the Sniper Rifles in Fortnite, was called "bolt-action."

With a bolt-action rifle, multiple bullets are loaded beneath the chamber, where they are pushed upward with a spring. When the bolt is pulled back, this spring pushes the topmost cartridge into the chamber. Then, the bolt is pushed forward again, which compresses the spring that is attached to the firing pin. Pulling the trigger releases a catch on this spring, sending the firing pin shooting forward into the primer at the back of the cartridge and firing a bullet. When the bolt is pulled back again after firing, another spring automatically ejects the empty shell casing while simultaneously loading another cartridge into the chamber. This allows the shooter to go through an entire magazine of cartridges while only needing to pull a bolt backward and then push it forward between each shot: much faster than having to pour in powder, drop in the musket ball, etc.

Between the bolt-action rifle, the revolver, and other similar spring-based firing mechanisms, for a long time most guns worked basically in this fashion.

The major turning point, and the one that really helps to explain what's going on with the rate of fire, was the invention of the machine gun.

The first machine guns were called recoil reloading machine guns, and they did an amazing job of taking advantage of Sir Isaac Newton's third law of motion, which tells us that every action has an equal and opposite reaction. When you fire a gun, the explosion of the gases inside the chamber propels the bullet forward out the barrel of the gun, while recoil pushes the gun back toward the shooter. So, with each shot fired, there is force going forward as well as force going backward. The recoil reloading machine gun takes advantage of this backward force to eject the spent shell casing while simultaneously pushing a fresh cartridge into the chamber.

To make this work, a spring is placed behind the bolt attached to the firing pin. Before the first shot, this bolt must be manually pulled back to compress the spring. Once the trigger is pulled, the spring releases, sending the firing pin into the primer and shooting the bullet. Then, the backward force of the recoil is allowed to push the bolt itself backward into the spring, using the force of the recoil to compress the spring so that the shooter does not have to do it manually.

Much like the bolt-action rifle, sending the bolt backward also opens up the chamber for a new bullet to enter. If the trigger remains in pulled position, the spring has nothing to keep it locked in a compressed position, so it automatically springs forward again and sends the firing pin into the next bullet. As long as the trigger remains pulled, the gun will continue to fire bullet after bullet.

Another common technology for making a gun fire automatically is called a gas reloading gun. As you recall, when a bullet is fired, the gunpowder in the cartridge ignites, expanding rapidly into a very large volume of gas, which then propels the bullet out of the barrel. In a gas reloading gun, some of this expanding gas is actually bled out of the barrel, directed down a special tube, and used to push the spring load and bolt backward, similar to a recoil reloading gun. Most modern machine guns and assault rifles employ a gas reloading system.

AUTOMATIC AND SEMIAUTOMATIC WEAPONS

What is the difference between a semiautomatic and an automatic weapon?

The simple, functional answer is that an automatic weapon requires only that the trigger be pulled for the weapon to keep firing, while a semiautomatic weapon requires that the trigger be pulled each time a bullet is fired. Both of these are different from a manual weapon, which requires some form of action to be taken by the shooter between each shot, whether pulling back the hammer on the revolver, a bolt on a hunting rifle, etc.

While there are many different types of technologies that make some guns automatic and others semiautomatic, most semiautomatic guns are simply automatic guns in which an extra pin has been added to stop the bolt from shooting forward after it is compressed back, until the trigger is pulled again. In this sense, most semiautomatic weapons actually employ extra functionality to stop themselves from being fully automatic.

Why do some automatic weapons fire faster than others, and why don't they just make all semiautomatic weapons automatic?

First and foremost, the decision to make a gun automatic versus semiautomatic is mostly made based on the intended uses of the weapon itself. Even apart from the fact that automatic weapons are generally illegal most places in the civilized world, there are many use-cases in which it is simply not ideal to shoot more than one round with each pull of the trigger. This can be due to the user wanting to limit the amount of damage being caused, or just as a means for conserving ammunition. This second reason is especially common: an automatic weapon obviously burns through ammunition much faster than a semiautomatic weapon does, and, unlike in Fortnite, real-world ammunition is heavy, expensive, and generally quite limited.

Similarly, the rate of fire of a fully automatic weapon is at least partially determined by the intended use of the weapon. The faster an automatic weapon fires, the more damage it does, and the more ammo it uses. Any situation where you want to do less damage and/or use less ammo will necessitate a slower rate of fire.

That said, from a technical standpoint, the rate of fire of an automatic weapon is also limited largely by the heat that the gun generates as it fires. Each time a bullet is fired, a small explosion takes place inside a gun, and 3,000-plus-degree gases are shot through the barrel. The higher the rate of fire of an automatic weapon, the hotter the barrel is going to get, and the less time it has to cool between each shot. Gun designers have found many ways to mediate this heat problem, including using air and/or water to cool the barrel, but no matter how it is dealt with, every gun will have a limit as to how much heat the barrel can handle from continuous automatic firing before the gun fails and stops working.

FIREARM ACCURACY

Why are some guns more accurate than others?

There are a number of reasons, especially as you increase the distance from the object at which you are firing. But to understand the science of why this happens, first we have to look at how a bullet manages to go straight in the first place. In the earliest guns, like the flintlocks, the bullet was a simple round ball that was fired through a smooth metal tube. While these guns would certainly do a lot of damage to anything they managed to hit, they were extremely inaccurate, usually missing their targets completely, especially when the target was more than a few feet away. This lack of accuracy was mostly due to the fact that the bullets simply did not fly in a straight line once they exited the barrel of the gun.

As soon as a bullet exits the barrel, many different forces immediately start to act upon it. The most influential forces are gravity, air resistance, and wind. Gravity exerts a constant force, pulling the bullet toward the ground. Air resistance works on a bullet in pretty much the same way it works on a person skydiving: every millimeter the bullet travels through the air is filled with lots and lots of tiny air molecules, and each one must be pushed aside by the bullet as it flies toward its target. The pushing aside of these air molecules slows the bullet down considerably while also making it tumble through the air, further throwing it off its course. Then there's wind: if you are firing a gun in the exact same direction that the wind is blowing, it won't have much of an effect on the trajectory of the bullet, but this is rarely the case. Instead, most of the time, the wind will be crossing the trajectory at an angle, constantly pushing the bullet as it does.

With the early smooth-barreled guns, these forces would instantly throw the bullet off its straight trajectory and send it careening in one direction or the other. Modern guns, however, employ what is known as the rifling inside the barrels to combat these forces. A rifled barrel has small grooves carved in a spiral running down the length of the barrel. These grooves are just large enough to spin the bullet as it passes through the barrel and exits the opening at the end, so that when the bullet starts to fly through the air, it is not only moving forward, but is also spinning around at the same time. This spin stabilizes the bullet's flight and allows it to travel in a significantly more consistent and straight direction.

How does this work? A spinning bullet travels straighter than a non-spinning bullet for the same reason that a spinning top will stay up as long as it is spinning while the non-spinning top will fall right over. In both cases, it is the gyroscopic effect of the spinning that gives the object its stability.

This can be a little bit hard to understand. Think about it this way: a bullet that isn't spinning is exerting force in only one direction: forward. So, let's say a gust of wind starts blowing exactly perpendicular to the path of the bullet. What will happen? Since the bullet is only pushing forward, there is nothing to push against the gust of wind coming from the side, so the bullet will be pushed out of its straight trajectory.

This is kind of like what happens if you are running forward and some-body comes and pushes you from the side. You'll probably lose your balance and fall over much more easily than you would if they came and pushed you from the front, because that is the way that your momentum is going.

However, with the spinning bullet, you not only have forward motion, but you also have motion going in all other directions perpendicular to your forward motion at the same time. Assuming a clockwise spin to the bullet, you will have motion pointing to the right at the top of the bullet, motion pointing down on the right side of the bullet, motion pointing to the left on the bottom of the bullet, and motion pointing up on the left side of the bullet (along with motion pointing in every other direction between them). As a result of having motion pushing in all of these directions, your bullet has momentum that can push against any force that might try to push it up, down, or to either side. As a result, your bullet can travel much farther and straighter than it would if it were not spinning.

In this same vein, the length of the barrel will also have a significant effect on the accuracy of the bullet once it is fired from the gun. The longer the bullet gets to stay in a forward-facing direction and gain spin from the rifling, the straighter it will be flying, and the greater its rate of spin will be when it finally exits the end of the barrel.

For example, consider the difference between a bullet traveling down the length of a twenty-four-inch-long sniper rifle barrel as opposed to a six-inch-long handgun barrel. With the short barrel, the bullet will only get to spin a few times before being flung out into the air, while the longer barrel will give the bullet four times as many spins. Not only that, but even if both guns were shooting the same caliber bullet, the larger quantity of powder in the sniper rifle cartridge will give the bullet a much greater velocity as it travels down the longer barrel. This combination of factors is like the difference between giving a top a really strong, tight spin, or giving it weak and wobbly spin. The stronger, tighter spin will allow the top to stay upright and spinning far longer than a weak and wobbly spin would. In this same way, the bullet fired at a higher speed through a longer barrel will fly far longer and straighter than a bullet fired at a slower speed through a shorter barrel. And of course, the straighter and faster the bullet is flying, the more accurate it will be over distance.

The last factor that affects the accuracy of a gun is the way that it handles recoil. This is especially important when talking about automatic weapons. As you recall, every time a gun is fired, due to Newton's first law of motion, an equal force is exerted backward toward the gun as is exerted forward with the bullet. This recoil force is distributed differently across the gun and the shooter with different kinds of weapons, and the more stable the gun can remain against the force of this recoil, the more accurate it will be. This is why a rifle with a butt that can be placed against the shooter's shoulder will always be more accurate than a handgun or a submachine gun that is held out in front of the shooter. Without having anything to stabilize the gun against your body, the force of the recoil will begin to push the barrel off of the spot where you were aiming it more and more as you continue to fire rounds. The same thing will happen with a full-sized automatic rifle; however, the greater stability gained by stabilizing the butt against the shooter's shoulder will allow this spread to be less noticeable while allowing the shooter to remain on target for more consecutive shots.

PART 3
Things That Go Boom

DYNAMITE

Okay, now we get to the really fun part: explosives. There are a number of explosive weapons and items in Fortnite, each with a different set of attributes, but they all share the basic and fundamental aspect of being explosive. But what is an explosion? What actually happens when you set off some dynamite, or throw a hand grenade, or launch a rocket with your rocket launcher? What makes something so small do so much damage?

To answer that question, we'll start by looking at the real OG that started the whole explosive thing. I am talking, of course, about dynamite.

What is dynamite exactly, and how does it work? It was invented in 1867 by Alfred Nobel, who would go on to make so much money selling it that he would wind up founding the Nobel Prize. Dynamite was the first stable and usable explosive material to be widely used after the invention of gunpowder. When Nobel invented dynamite, however, it wasn't by figuring out how to make a new substance explode; rather, his great innovation was figuring out a way to make the explosive chemical in dynamite *not* explode. Allow me to explain.

Twenty years before Alfred Nobel invented dynamite, an Italian chemist named Ascanio Sobrero was experimenting with different compounds of nitrogen when he came across a new chemical called nitroglycerin. Nitroglycerin molecules contain three atoms of carbon, five of hydrogen, and nine of oxygen. Immediately after creating this new substance, Sobrero realized it was quite unique. Nitroglycerin is a type of chemical known as a contact explosive (and an extremely unstable one at that). This means that, in its pure form, any physical shock will cause nitroglycerin to explode. Hit some nitroglycerin with a hammer? *Boom!* Drop some nitroglycerin on the floor? *Boom!* Put some nitroglycerin in your back pocket and try to do the Carlton dance? *Extra boom!*

Pretty much any significant physical contact with nitroglycerin, or any significant amount of movement, will cause it to explode. It was so unstable and explosive that its inventor didn't actually think there could ever be a use for it, as it was simply too dangerous to make or transport anywhere.

What Nobel discovered twenty years later was that by mixing nitroglycerin with a dry, absorbent material (he used a mixture of shells and baking soda, but even something like sawdust would work), you can create a clay-like putty that can be formed into any shape you require, and does not explode simply by shaking it or dropping it. Believe it or not, at the time that Alfred Nobel was working on his new invention, the biggest reason that people needed explosives was not actually to blow up man-made forts on top of temporarily dormant volcanoes. Rather, the biggest thing people wanted to use explosives for was mining: blowing up large chunks of rock and earth to get to the rare and precious metals beneath.

With Nobel's new invention, one could easily drill a hole in a large slab of rock, then shape this explosive substance into a long skinny cylindrical stick, wrapped in red wax paper, that was perfectly sized to fit inside of the hole. Also, unlike pure nitroglycerin, dynamite required the use of a blasting cap (basically a small quantity of gunpowder ignited with either a lit fuse or a percussion cap) to detonate. This meant that as long as the miners waited until the last second before attaching the blasting cap, they could safely transport and set up the dynamite without having to fear an unwanted explosion.

So then, what exactly happens when you detonate a stick of dynamite? Why does nitroglycerin explode in the first place, and why does that make dynamite explode as well? What is actually happening in that little red stick?

Like any explosion, a nitroglycerin-based explosion is, at its most basic and simplistic level, an extremely rapid and violent expansion of gases. If you think about how gunpowder works in a cannon or a bullet by very rapidly expanding from a solid state into a much larger gaseous state, you have the right general idea, though the chemical properties of nitroglycerin are significantly different than gunpowder, and give it some unique characteristics.

Nitroglycerin's extreme instability means that the chemical bonds that hold the different parts of the nitroglycerin molecule together can be easily broken apart. Breaking the chemical bonds that hold any molecule together causes the elements that make up that molecule to either remain separate and/or form new chemical compounds.

For any combustion to take place, you need three things: a source of fuel, an oxidizing agent, and something to start a chain reaction and get it all going (otherwise known as the detonator). For example, let's say you filled a bucket with gasoline (please don't actually do this): what would you have? Fuel. The next thing you need is an oxidizing agent. Well, considering that your bucket of fuel just happens to be on the planet called Earth, and Earth has an atmosphere that is positively filled with oxygen (one of the best oxidizing agents there is), you're all set with that piece, too. Then you just need something to get your reaction started, so you take a match and throw it into the bucket of gasoline. Fuel, oxidizing agent, detonation: *boom*.

A quick note about oxidizing agents. This performs a very simple task during combustion: it disturbs the electrons orbiting the atoms that make up the molecules of the fuel source, causing them to break apart. As these molecules are also made up of atoms that have electrons orbiting them, pretty soon you have electrons flying around breaking apart lots of molecules. This breaking apart is what we think of as combustion, because when the molecules break apart, they create heat and form new chemical bonds.

In the case of nitroglycerin, the chemical bonds are weak in the first place, so they are easily shattered by simply striking them with something like a hammer. When these first nitroglycerin molecules are broken apart, the oxidizing agent combines with the fuel in each molecule to create new molecules that are gaseous and take up more space in the original molecules. Additionally, because you're breaking apart oxidizing agents inside the molecules, each time a molecule comes apart, it instantly shatters all the other molecules near it. This creates a chain reaction that almost instantly causes every nitroglycerin molecule to break apart, producing a large volume of gas and heat in the process, and not stopping until every nitroglycerin molecule in direct contact has been obliterated. This molecular process turns a small piece of cool nitroglycerin into a very large cloud of hot gases.

Let's look at a step-by-step breakdown of what happens when this process plays itself out with nitroglycerin.

- **Step 1:** You see a big old pile of nitroglycerin just sitting around on your living room floor, conveniently right next to a nice, shiny hammer. Being the destructive little weirdo that you are (this is an assumption, of course, but being that you play enough Fortnite to be reading this book, it's probably a safe one), you decide to pick up the hammer and strike that pile of nitroglycerin just as hard as you can (again, please, please do not ever do this).
- **Step 2:** As this is pure nitroglycerin, the force of the hammer coming in contact with the nitroglycerin molecules on the surface of the pile causes all of the molecules that absorb the shock of this strike to begin breaking apart as their chemical bonds shatter from the impact.
- **Step 3:** As you know, for any type of combustion to take place, you need fuel to act as a source of energy and an oxidizing agent to release that energy from the fuel. What you may not have realized is that each and every nitroglycerin molecule has both of these components built right into it: fuel and oxygen. Each nitroglycerin molecule is made up of a single hydrocarbon fragment (a particularly energy-rich fuel source) attached to three nitrate groups (very powerful oxidizing agents). Upon striking the nitroglycerin, these four pieces are instantly freed from their chemical bonds, whereupon they sort of slam into each other (not really, but it's a good way of picturing the process) and create a tiny one-molecule-sized explosion.
- **Step 4:** This tiny one-molecule-sized explosion at first just occurs in the molecules that absorbed the shock of the hammer strike. However, because nitroglycerin is so unstable, this first round of little explosions causes all of the molecules *next to and around* the first set of molecules to also break apart and explode.

- **Step 5:** Each nitroglycerin molecule that explodes winds up detonating many, many more molecules. No one knows the exact number of molecules destroyed by each exploding neighbor, but for the sake of an example, let's assume it's 100. So, if the hammer strike were to cause 100 molecules to detonate, those molecules would then cause 10,000 more molecules to detonate, which in turn would cause 1,000,000 more to detonate, and then 100,000,000, and then 10,000,000,000, and on and on until all of the nitroglycerin molecules have been destroyed. This shockwave of detonating molecules makes its way through all of the available nitroglycerin at a speed thirty times greater than the speed of sound.
- **Step 6:** As each nitroglycerin molecule detonates, its component atoms instantly begin to rearrange and form new chemical compounds like N_2 and CO, which have stronger, more stable chemical bonds than nitroglycerin. This process of rearranging atoms generates so much heat that the resulting chemical mixture winds up having a temperature of about 5,000°C (9,000°F).
- **Step 7:** Here's the real juicy part of this whole process: these brand-new, super-heated molecules take up almost 1,200 times as much physical space as the original nitroglycerin molecules did. This means that, faster than you can blink an eye, your little innocent-looking pile of room-temperature nitroglycerin will become 1,200 times larger than when this all started, while simultaneously heating up to 5,000°C. This is the actual explosion.

What does this have to do with dynamite, exactly? Well, as noted earlier, Alfred Nobel's great innovation was to make this chemical reaction more difficult to start. By combining the nitroglycerin with inert substances, you are essentially stabilizing the molecules in a nice, cozy cushion that requires a bit more of a punch to get the whole reaction started. That's dynamite: nitroglycerin that has been combined with other substances to make it more difficult to detonate, and therefore stable enough to use in actual, practical applications.

Of course, dynamite is just one of many different types of chemical explosives, and while the chemical structures of all such explosives are unique, the basic science behind how they all function is the same: you're taking a small, relatively cool substance and creating a chain reaction on a molecular level that turns it into a large volume of hot gas.

HAND GRENADES

s you probably know from your extensive video gaming experience, while dynamite may be quite useful for explaining some of the basic science of how explosives work, it does not necessarily make a particularly good weapon, and it is much more suited to damaging buildings than people. For a more useful and ubiquitous explosive weapon, you need not look any farther than the trusty hand grenade.

The hand grenade is an explosive that, of course, fits in your hand, and one that you can easily throw to create a moderate-sized explosion. It doesn't seem that different from dynamite, does it? It's actually quite a bit different. First of all, as stable as dynamite is in comparison to pure nitroglycerin, it is not really stable enough to be worn on the belt of a combat soldier. The nitroglycerin inside the dynamite is simply too unstable for that kind of use. Instead, during World War I, the British created a hand grenade that used TNT as its primary explosive material.

TNT stands for trinitrotoluene and has a chemical formula of $C_6H_2(NO_2)_3CH_3$. You will notice right off the bat that those letters are quite similar to the ones found in nitroglycerin, $C_3H_5N_3O_9$, but in a slightly different order. In fact, the elemental components of TNT are nearly identical to those in nitroglycerin; however, the chemical bonds that hold TNT's elements together are far stronger. As a result, TNT, though not quite as powerfully explosive as nitroglycerin, is far more stable. It is so stable that it can even be melted down into a liquid and poured into metal casings (a process that would not be recommended with nitroglycerin). This makes TNT the ideal explosive for hand grenades.

Aside from containing a more stable explosive, what makes a hand grenade different from a stick of dynamite? The main thing is the way its mechanical components function together to create a specific kind of explosion that make it optimal for a very specific set of purposes.

There are six main parts to a standard hand grenade: the body, the safety pin, the safety lever, the delay fuse, the detonator, and the charge. Before it is used, the pin of a hand grenade is fit into a hole on the spring mounted safety lever, keeping the lever in place. When you want to use the grenade, you must first hold tightly onto this lever and pull the pin. It's important to hold onto the lever to keep the spring from popping it up before you are ready to throw the grenade.

When you do throw the grenade, this spring pops up the handle, and the detonation process is set in motion. The mechanical popping action of the lever ignites a delayed fuse inside a metal tube that travels down toward the center of the grenade. These fuses are specially made so that they take a precise amount of time to burn. Usually, this is about four seconds: just long enough for the grenade to reach its target. Once this fuse has burned all the way down to the bottom, it ignites a detonator, which is essentially a small explosive chemical that ignites as soon as the lit fuse touches it. The detonator sends out a shockwave throughout the grenade, which in turn causes the TNT charge to explode.

At this time, a chemical reaction happens inside the hand grenade that is very similar to nitroglycerin: once the chemical bonds of the TNT are broken apart, the oxidizing agent inside the chemical reacts with the carbon-based fuel inside the chemical to turn a small, cool volume of TNT into an extremely large and hot volume of gas.

What really makes a hand grenade a hand grenade, however, is not the explosion itself, but rather what the explosion does to the grenade *casing*. The earliest hand grenades looked like little metal pineapples for a very good reason, and it had nothing to do with SpongeBob. Each of the little sections of the pineapple-shaped piece of metal were separated by thinner pieces of metal, so that when the grenade exploded, the casing broke apart right along these gridlines. This made each little raised section of the grenade separate from the others and shoot out in every direction with the force of the explosion, thus turning this little metal ball into dozens of bullets flying in every direction at the same time. For the most part, it is the damage caused by these flying pieces of metal that is responsible for the destruction a grenade can cause, not the explosion itself. Explosions in a grenade, like the explosions inside of bullets, are meant primarily to push the little pieces of metal fast enough and far enough so that they go through people's bodies and kill them.

In Fortnite, this is clearly not how the grenade delivers damage. If that were the case, any player within a wide range of the grenade would receive damage when it explodes, similar to what a player would receive from small-caliber bullets (like those shot from a submachine gun, for example). The explosive force damage, meanwhile, would be high, instantaneous, and similar to dynamite, though very much localized in the immediate vicinity of the explosion. While we do clearly see the Fortnite hand grenades dealing this kind of explosive damage, we don't see any of the bullet-type damage indicative of flying shrapnel. From this, we can deduce that there is no shrapnel coming out of the grenades in Fortnite, and that they are meant simply as explosive devices used for their concussive force (which probably means that they are better suited to destroying structures and vehicles than eliminating other players).

GRENADE LAUNCHERS

I f you're anything like me, there is something that is just so satisfying about the loud, sonorous *thwonk . . . BOOM* sound that the grenade launcher makes when it volleys a grenade, lets it arc for a few seconds in the air, and then explodes. I honestly don't know if I've ever managed to get an elimination with a grenade launcher, but it's just so darn satisfying that I find it impossible to resist whenever I come across one in the game.

So, what is the science behind the grenade launcher? How does a grenade launcher work in the real world, and is it anything like the way it works in Fortnite?

At its most basic level, a grenade launcher takes the scientific principles that make both grenades and guns work and combines them. This should be pretty obvious just by firing off a few rounds, but there are actually a few more interesting things going on there as well.

There have been many variations on the grenade launcher through the years, but the one you find in Fortnite is known as a revolver-style grenade launcher, similar to the Melchor N32 from South Africa. This style is far and away the most popular currently in use by today's military, as it is simple, effective, and easy to use. The N32 (and others like it) grenade launcher is capable of carrying six 40 mm grenade cartridges and has a range of about 100 yards.

The first thing you need to know about this kind of grenade launcher is that the 40 mm "grenades" it fires are quite different from your standard hand grenade (some people have actually made the argument that these rounds are not really grenades at all, but I think that's just splitting hairs). Pretty much the only aspect of the 40 mm grenade launcher round that is similar to the standard hand grenade previously discussed is that they both use explosive charges made from TNT, or another similar chemical explosive.

Okay, so what about the differences? First, there is no pin nor safety lever on the 40 mm grenade round. This is because most grenade rounds are designed to explode on impact, as opposed to after a specific interval of time (there are hand grenades that explode on impact, but these are relatively rare). In typical grenade launchers of this style, the impact detonation is accomplished with a small empty space at the very tip of the grenade that becomes compressed when the grenade hits its target. This compression sets off a fuse, which instantly detonates the TNT charge in the grenade.

Also, and perhaps most obviously, this little explosive grenade is mounted on the front of something like a bullet cartridge. This cartridge does not, however, simply ignite a bunch of gunpowder right next to the grenade, as that would most likely just make the charge explode right there in the shooter's hands. Instead, these rounds have a small pocket of combustible chemicals (similar to gunpowder, but with a slightly different chemical composition) that, when ignited, expands its gases into an expansion chamber inside the cartridge and out the back of it at the same time. This has the beneficial effect of decreasing the shock on the grenade charge in the tip so that it doesn't detonate before it reaches its target. Unfortunately, this benefit comes at a cost: the grenade launcher's range is far shorter than any standard bullet.

More modern 40 mm grenade rounds even have a centrifugal-force-operated safety switch inside them that only lines up the detonator with the charge after the cartridge has completed a specific number of spins at a high velocity. The spinning of the grenade (which, as you know, is necessary for it to fly straight) forces the moving part of the switch toward the outside of the shell via centrifugal force, lining up the detonator with the charge and preparing the grenade for detonation. The purpose of this device is to allow the grenade to detonate only after traveling a safe distance away from the shooter. So, for example, if you were to shoot a grenade at an enemy combatant, but then that combatant were to instantaneously build a wall between the two of you so that the grenade bounced back toward you, the round would not have spun enough to activate the centrifugal switch, and it would not detonate, and you would not die.

As we know, this is not the kind of grenade launcher round used in Fortnite. Even though the Fortnite grenade launcher does look an awful lot like that Melchor N32, it is different in a few significant ways, most obviously that its rounds are not impact detonated at all, but rather seem to utilize a time-delayed fuse of some sort so they go off at a precise interval after being fired. It is possible to use this type of round with a real grenade launcher, but they are quite uncommon.

ROCKET LAUNCHERS

I f you have played even a little bit of Fortnite (and I highly doubt you'd make it this far into this book if you hadn't) you know that in the vast majority of circumstances, the rocket launcher is a much better weapon than the grenade launcher (yes, I know the grenade launcher can be effective in certain situations when you're part of a squad, but come on—that's pretty rare). The rocket launcher fires its explosive very far, very fast, and very accurately, while the grenade launcher simply lobs its explosive gingerly in the general direction you point it. Sure, the grenade launcher, being a revolver-style weapon, can fire six rounds in the time it takes the rocket launcher to fire two, but this hardly makes up for the grenade launcher's other issues.

What makes the rocket launcher able to fire so much more accurately over such a greater distance than the grenade launcher? Is there anything else about the rocket launcher that's different from the grenade launcher? Also, is the rocket launcher in Fortnite anything like rocket launchers in the real world?

To answer these questions, we first need to define what exactly we mean by a rocket launcher. Fortnite's rocket launcher is actually quite similar (at least from the outside) to a real-life weapon, but the real-life weapon is not called a rocket launcher: it's called a rocket-propelled grenade launcher. Furthermore, the Fortnite rocket launcher is essentially the same design as the Soviet-designed rocket-propelled grenade launcher known as the RPG-7. First designed during the height of the Cold War in the early 1960s, the RPG-7 utilized German technology from World War II to create a shoulder-mounted mortar-launching device that would be reliable, easy to use, and devastatingly destructive. At the time, the Americans were using the famous bazooka as a similar device, but it was never as reliable or popular as the RPG-7.

So right there in the name, you can see a clear similarity between the grenade launcher and the rocket launcher: they are essentially different ways of firing a small explosive farther than one can by simply throwing it. The main difference is in the technology they use to propel the explosive out of the tube and toward the target. As discussed above, the grenade launcher utilizes a small, fairly simple explosive charge that can only launch the grenade a relatively short distance at relatively slow speeds. The RPG launcher, on the other hand, uses different technology entirely.

Now, nearly all of this technology is in the RPG round, but before we get to that, let's look at the launcher itself. A rocket-propelled grenade launcher is not a very complicated device, being essentially just a long tube with a trigger mechanism, sites, and a breach at the back. A breach on an RPG launcher is a conical hole at the rear of the tube that allows escaping gas from the rocket-propelled grenade to exit the launcher far enough behind the shooter so as not to cause any harm (the exhaust from a rocket is quite hot, and would easily burn you if you were standing right behind it, or if the exhaust vented onto your shoulder instead of behind you).

What really makes a rocket-propelled grenade unique is its two-stage propulsion system. This consists of a booster for getting the RPG out of the launcher, and a sustaining motor that ignites after the RPG has made it a short distance in front of the launcher. It is this sustaining motor that actually propels the grenade toward its target. This two-stage process is vital, as it allows the powerful sustaining motor to use highly combustible propellants that are far too hot to be ignited inside any kind of handheld launcher (without killing the person holding it, anyway).

This is what happens when you fire a rocket-propelled grenade:

1. Pulling the trigger sends a spring mounted firing pin into the primer of the RPG, which, just like with a grenade launcher or a bullet, ignites a small powder charge.
2. The small powder charge violently expands into a large volume of hot gas, which propels the RPG out the front of the launcher (and, because even rocket launchers have to follow Newton's laws of motion, exerts an equal amount of force with the gas and the recoil going backward behind the launcher). This relatively small

powder charge is powerful enough to push the rocket out of the muzzle of the launcher at about 120 meters per second without creating so much exhaust that it winds up melting the shooter and everyone else in their immediate vicinity.

3. This is when things get really interesting. There is a device inside the RPG's propulsion system called a piezoelectric fuse, which utilizes some interesting properties of certain types of crystals that actually create an electric charge when acted upon by certain physical forces. In this case, the G-forces that are created as the rocket accelerates cause the piezoelectric fuse to release an electric charge into a gunpowder primer.

4. Once ignited, the gunpowder primer in turn ignites a small quantity of nitroglycerin (remember, the thing inside dynamite?), which then ignites the rocket propulsion system.

5. The rocket propulsion system may use any one of a number of different types of chemical propellants, but all of them work more or less the same: they burn very hot at a very specific rate, creating large amounts of gas that steadily propel the rocket forward for a period of time determined by the total quantity of propellant inside the sustaining motor. As with all rockets, this forward propulsion is achieved by directing the flow of expanding gases backward so that the trajectory of the grenade continues forward. It is important to note here that the direction of the rocket fuel exhaust gas is usually pointed at a slight angle, and expelled through a few different nozzles, so as to allow the rocket-propelled grenade it to spin as it makes its way through the air. Unlike a bullet or a grenade launcher cartridge, there is no rifling inside the barrel of a rocket launcher, so in order to create the spin necessary to maintain a straight and accurate forward trajectory, the spin must be achieved by directing the flow of rocket exhaust in a slightly spiraling motion.

6. Once the sustaining motor kicks in, which happens approximately eleven meters in front of the barrel of the launcher, the speed of the rocket-propelled grenade increases to its maximum velocity of around 300 meters per second.

The actual grenade at the front of an RPG (also sometimes referred to as its warhead) is quite similar to the explosive part of most grenade launcher rounds: it is essentially a fuse mounted on the tip of the explosive that, when it hits an object, ignites the explosive charge inside the grenade. The charge itself can be made out of a number of different types of explosives (depending on how the RPG is meant to be used), but regardless of which explosive is utilized, it basically functions similarly to the way that dynamite or TNT functions: a detonator creates a chain reaction at the molecular level that causes a relatively small and cool volume of solid material to become a very large and very hot volume of gas, destroying anything that happens to be near it when it does.

So, how does this compare to the way the rocket launcher works in Fortnite?

The Fortnite version seems to work quite similarly to the real-life version, with only a few small exceptions. Probably the clearest and most noticeable difference is the range and accuracy versus a real RPG launcher like the RPG-7. In Fortnite, when you shoot a rocket from your rocket launcher, it pretty much hits dead-on its target every single time, as long as that target is within the rocket launcher's range (more on that shortly). You can be just as accurate at 300 meters with the rocket launcher in Fortnite as you can at 10 meters.

This kind of accuracy is clearly not very realistic. According to the US Army Manual, at more than 200 meters the RPG is only accurate 50 percent of the time. This goes down to 10 percent when you get to 300 meters. Beyond that, I don't think anybody would suggest shooting an RPG at all.

This brings us to the overall range. While some rocket-propelled grenade launchers will just keep on flying until they run out of propellant or hit something, most do not work that way anymore. Nearly all RPGs currently in production have a maximum range, after which the explosive will detonate whether it has hit a target or not. This range is usually set at about 950 meters, in part because it is so difficult to be accurate at this range. Why allow a deadly explosive to just keep flying when you have no idea where it's going to land? The rocket launcher in Fortnite also detonates after a set distance, but that distance is much shorter: about 300 meters, give or take.

Another obvious difference between an RPG launcher and the Fortnite rocket launcher is the speed at which the rocket flies through the air. Watching a video of an actual RPG-7 getting fired on an episode of *MythBusters*, I noticed that the rocket seems to fly through the air and reach its target almost instantaneously because a typical RPG has a muzzle velocity of around 120 meters per second, which after only eleven meters of flight increases to nearly 300 meters per second when the sustaining motor kicks in. This means that it would take a real RPG approximately 1.2 seconds to reach a target 300 meters away.

Knowing this, I set up an experiment in Fortnite. First I found a nice high mesa in the desert with a pretty little cactus on top. I then placed my marker on that exact spot and built a tower in the valley beside it, positioned so that when I reached the same height as the mesa I would be exactly 300 meters away from the cactus. Then I simply fired my rocket launcher at the cactus and timed how long it took for the rocket to reach its target (actually, I recorded the whole thing, then went back and timed it later, because I wasn't quite quick enough with the stopwatch to do it live). Amazingly, this is how long it took: twelve seconds!

That's right, the rocket in Fortnite travels at an average speed of about twenty-five meters per second—less than a quarter of a real RPG's muzzle velocity, and less than 10 percent of the speed a real RPG goes with its sustaining motor. Why did Epic decide to make the rocket so much slower in the game then in real life? Frankly, I don't know. Maybe it just makes for a more balanced weapon, maybe they wanted to give the player being shot at a chance to get away, or maybe it's just fun to watch that trail of smoke billow away from you as you wait for your enemy to meet his demise.

QUAD LAUNCHERS

We're not going to spend too much time discussing the Quad Launcher here, because it just isn't that unique a weapon in Fortnite. It functions in the game as a sort of hybrid between the grenade launcher and the rocket launcher. It carries more rounds than the rocket launcher (four, as the name suggests, but not as many as the six rounds that the grenade launcher carries). It can shoot a little farther than the grenade launcher, especially if you point it up at an angle, but it can't shoot nearly as far as the rocket launcher. As far as damage is concerned, it has a DPS on par with the other two, though it's a little less powerful than both.

However, as with the grenade launcher and rocket launcher, the Quad Launcher does seem to be very similar to a real-life weapon: the M202 incendiary rocket launcher, developed by the United States during the Vietnam War. The M202 looks more or less identical to the Quad Launcher in the game, even including the fold-down cap on the front. It also takes a four-cartridge clip of rockets.

The real M202 was never made with explosive rounds like the ones used in the game, which explode very similarly to the rounds used in both the rocket launcher and the grenade launcher. Instead, the M202 fired incendiary rockets, and was used by the US Army as a replacement for the World War II flamethrower.

Basically, when the M202's round hits its target, instead of sending out a shockwave of hot gas that blows up anything in its immediate vicinity, it spreads a chemical in liquid form that sticks to anything it touches and begins to burn as soon as it comes in contact with the air.

It is also worth noting that the M202, while it does use rockets that look rather similar to those used by the rocket launcher, does not have a booster rocket like the RPG-7 does. Instead, when you fire this type of quad rocket launcher, the round is propelled by a single solid fueled rocket motor that works very much like the hobby rockets you might fire in your backyard, in which a solid fuel is ignited and burns very fast for a short period of time, propelling the projectile forward.

THE STINK BOMB

Ah, the Stink Bomb. That little blue ubiquitous grenade that you probably pass over a dozen times in every round in favor of more powerful throwable weapons. In Fortnite, it seems almost whimsical: akin to the stink bombs that you can buy at magic shops and dime stores around the world that are merely little drops of nasty-smelling sulfurous liquid in a thin glass container.

However, a quick look at real-world stink bombs makes you realize that it is far more insidious than that. So insidious, in fact, that there have been treaties banning its use going as far back as Article 23 of the Hague Convention of 1899.

If the Stink Bomb existed in the real world, there would only be one name for it: a chemical weapon, and a nasty one at that.

Throwing the Stink Bomb in Fortnite does not cause any physical damage to structures, weapons, or vehicles, therefore it doesn't affect the player's shield (and isn't blocked by it, either). Instead, when the Stink Bomb hits its target, it creates a ten-meter-diameter sphere of greenish gas that causes ten points of damage per second. Thankfully (if you are the target of the Stink Bomb), there is only enough gas in a single Stink Bomb to last for nine seconds, so it won't kill you if you're at full health. Regardless: we are talking about a gas bomb that can kill a person in ten seconds.

What do we call a gas that can kill you in ten seconds? Does such a thing even exist?

Any weaponized form of gas that can cause fatality in humans is certainly some kind of chemical weapon, and, as such, one of the most dangerous devices humanity has ever created. If you know your history (or even your current events), you know that the use of weapons like this is completely illegal in virtually every country on the planet, as well as under any conventions of war or international treaties.

You might be thinking: wait, out of all the weapons in Fortnite, this is the one that is actually the worst in real life? It barely even does any damage!

While it may be true that the Stink Bomb is not the deadliest weapon in Fortnite, that's because it only comes in such small quantities. You can imagine what would happen if, for example, you could create an extraordinarily large Stink Bomb that was capable of, say, covering all of Neo Tilted (or the entire island) in a cloud of green gas for multiple minutes. It would be even more deadly than the Storm (which takes a whole lot longer than ten seconds to kill you). At this scale, there is another word for the stink bomb: weapon of mass destruction. If you made a stink bomb in real life, and made it large enough to cover a whole city, you would have a weapon as deadly as anything ever created by humankind, capable of killing millions of people in less than a minute.

So if the Fortnite Stink Bomb is in fact a chemical weapon, what kind is it?

Actual chemical weapons are generally called chemical agents, and there are quite a few different types of them out there. To figure out exactly which chemical agent is in a Fortnite Stink Bomb (or at least the one with the most similar properties), we need to start by listing everything we know about the Stink Bomb, and then seeing if any of the known chemical agents meet that description.

1. The chemical agent in the Fortnite Stink Bomb is a gas or a vapor of some kind.

This fact, while helpful, only narrows down our search a little bit. Most of the common chemical agents used in chemical weapons can be made into a vapor or gas, although not all of them are most effective in this state. For example, VX, one of the deadliest chemical weapons on Earth, kills its victims mostly by dropping into a liquid and getting on people's skin. That said, considering that almost any chemical agent (or at least any of the broad categories of chemical agents) can be made into a gas more or less like the kind that comes out of the Fortnite Stink Bomb, we can't really narrow anything down just yet.

2: The Fortnite Stink Bomb is stinky.

Okay, I don't really know what the Fortnite Stink Bomb *actually* smells like because Fortnite is, you know, a video game, and you can't smell video games (at least, not yet). Nonetheless, I think it is safe to assume that the Stink Bomb is not ironically named, and is actually at least a little bit smelly.

So do all chemical agents smell bad? Do any chemical agents smell good? Can we narrow down our search at all with this information?

These are actually rather difficult questions to answer. First of all, whether something smells good or bad is fundamentally a qualitative judgment and a matter of personal opinion. Something that smells perfectly fine to me might smell quite disagreeable to you, and vice versa. Also, seeing as how forming a subjective opinion requires actually experiencing the smell of something, and nobody wants to go around smelling chemical agents and deciding if they like their odor or not, there really isn't a lot of information out there about how people have judged these particular odors in the past. So we'll have to set aside the question of which chemical agents smell "good" and which ones smell "bad." That said, we can figure out which chemical agents are considered odorless, and which ones are known to have a distinct odor. This information, being very useful in the detection of such agents, is widely known and readily available. Thus, we can eliminate any of the chemical agents that are actually odorless. After all, something that is odorless would probably not inspire the name "Stink Bomb," would it?

So, which chemical agents can we eliminate based on whether they have an odor? Quite a few, actually. There's arsine, VX, and lewisite, for starters. Sarin is considered odorless in its pure form, but it is rarely found in its pure form, so that one is kind of a wash. In the end, while we are able to eliminate about 30 percent of known chemical agents, we can't eliminate any one entire category yet, so let's just keep going and see what we come up with.

3. The gas inside the Fortnite Stink Bomb is fatal in ten seconds.

Here is where we can really start to narrow it down a bit. While many of the remaining chemical agents are fatal in high enough concentrations (and with long enough exposure), only a few of them can kill someone in under a minute, with most of them taking anywhere from an hour to multiple days to cause fatality.

Now we can eliminate a few of the broad categories of chemical agents. There are basically four categories: blistering agents (like lewisite and mustard gas), nerve agents (like sarin, soman, and VX), blood agents (like cyanogen chloride and hydrogen cyanide), and choking agents (like phosgene, diphosgene, and chlorine).

As it turns out, pretty much all of these agents work very much as their names suggest:

- *Blistering agents:* These cause skin irritations and blisters that are basically severe chemical burns. Furthermore, when these types of chemicals are deployed as a weapon in a gaseous form (like mustard gas), they can cause chemical burn blisters inside the respiratory tract, causing a great deal of damage. Mustard gas, being a yellow-green gas with a distinct odor, could be a good contender for our Stink Bomb, if it weren't for the fact that its effects take multiple hours (or even days) to cause fatality. In fact, pretty much all of the common blistering agents take too long to work for them to really make sense as a possibility for the Fortnite Stink Bomb. So, we can move along to the next category.

- *Nerve agents:* If you guessed that these chemicals affect your nervous system, you are correct. Pretty much all nerve agents function in a similar fashion—by inhibiting a certain enzyme that breaks down the neurotransmitter acetylcholine, damaging the synapses that control muscle contraction. This causes just about every muscle in the body to contract simultaneously and lose any ability to relax at all. If this sounds painful to you, it certainly is—especially considering that muscles like your heart and lungs are just as affected as every other muscle in your body. The end result of exposure to a nerve agent is that you are completely paralyzed, inside and out, and usually die within a few minutes. Considering the speed at which fatality can occur, nerve agents are still certainly a contender for our Stink Bomb.

- *Blood agents:* These affect—no surprise here—the blood. Blood agents can be ingested in multiple ways, though they are usually dispersed in some kind of aerosol or gas. They tend to all have similar effects on the human body, working by blocking the body's ability to absorb oxygen from the blood. The main reason we have blood at all is to carry oxygen around our body and deliver it to all of our different types of cells, so when you take away the blood's ability to deliver that oxygen, pretty much every cell in your body starts to suffocate almost immediately. Similar to nerve agents, blood agents can take effect within seconds or minutes, so we can keep these on the board as well.

- *Choking agents:* Also known as pulmonary agents, this is the category of chemical weapons that, well, cause their victims to choke to death. Though there is a fair amount of variety on this category, in one way or another each of the choking agent chemicals inhibits the victim's ability to breathe, causing imminent death. That said, while the basic way these chemicals affect the human anatomy is the same, the timing and chemical type can be very different. While some can take 24 to 48 hours to cause fatality after exposure, there are some that kill in a matter of minutes. The latter type can stay on the board too.

So what are we left with? After looking at each of the four major categories of chemical agents, we can rule out the blistering agents and limit our search to nerve agents, blood agents, and any choking agents that work particularly quickly.

4. Stink Bombs create a greenish, translucent gas.
Here is where I think we can finally figure out which chemical agent is our winner. Now that we have narrowed our search down to nerve agents, blood agents, and a few choking agents, we just need to see which (if any) of these can be made into a greenish, translucent gas.

Right away we can eliminate all of the major blood agents. Phosgene, arsine, cyanogen, and all of the rest of the blood agents are all colorless in their gaseous forms, and thus not the right pick for the Fortnite Stink Bomb.

Nerve agents, it turns out, are also pretty much all colorless, and most of them are actually odorless as well. In any case, out of all of the few nerve agents that are spreadable in a gas and can be lethal in under a minute, none of them are green and/or stinky.

This brings us the choking agents, and a quick review of the color and order of each of these shows that only one is a real contender for the Fortnite Stink Bomb: good old-fashioned chlorine. That's right, the same chemical that you use to get the skid marks out of your shorts, in its gaseous form, is a greenish-yellow, very unpleasant-smelling gas that, in high enough doses, can be fatal very quickly.

All that said, is it really possible to make a stink bomb that is as fatal as the ones in Fortnite? Probably not. Even with chlorine gas, you would need to get the concentrations in the air to exceed at least 1,000 parts per million (though you would probably need to be closer to 10,000 parts per million) to achieve death in under a few minutes. Even if you were to inject the chlorine gas directly into somebody's lungs, it is doubtful that they would die in ten seconds. More importantly, making any kind of grenade-type device out of chlorine gas that could create concentrations of gas outdoors, in the open air, would probably be impossible.

THE SHADOW BOMB

A h, the Fortnite Shadow Bomb: Epic's wildest Season 8 addition to their arsenal of weird-ass grenades. The Shadow Bomb, as you know, looks sort of like a regular grenade: it's a black cylinder with a pin and a lever, much like any old hand grenade. But when you throw the Shadow Bomb, something very different happens: mainly, you turn temporarily invisible. For about six seconds, your physical form disappears, and then slowly reappears as wisps of gray smoke before you finally become whole again. At the same time, while you are invisible, you have the ability to jump off walls, and even double jump in the air.

On the surface, this might seem like it belongs in the same category as the Boogie Bomb: a whimsical, fantastically magical, purely fictional item, with no bearing in scientific reality whatsoever. And while it is true the technology to create an actual shadow bomb certainly does not exist (spoiler alert), the science of invisibility is very real indeed.

It may seem like pure science fiction, but scientists have been working for a long time, and with a good deal of success, at the task of making objects invisible. But how are they doing it? What are the scientific concepts that may someday allow scientists to make things, like people or even vehicles, invisible?

To explain that, we have to start by looking at how you see things in the first place. Let's say you have a banana sitting on the table in front of you in a completely sealed room with no light whatsoever. Can you see the banana? No, of course not. You can't see anything, because it is completely dark inside this room. Now let's say we take a small flashlight and shine it from the very top of the room onto the banana. You will now see the banana sitting on the table quite clearly. But why?

It's quite simple. Imagine light as a series of tiny lines or rays in different colors being shot out of the light source toward the banana. When all these different colors of light rays hit the banana, some are absorbed and others begin to bounce off in all directions. Specifically for a banana, it is only the yellow light rays that are reflected. Bananas are made out of a substance that reflects the yellow light. That's why the banana looks yellow to you: all of these yellow light rays bounce off the banana and hit you in your eyes, at which point the nerves in your eyes absorb these yellow light rays and send messages to your brain, telling it what color the banana is.

Let's stay in this room and keep our little light on, but imagine that I really, really don't want you to see the banana. How could I go about hiding it?

Well, if the banana is absorbing every color except yellow, couldn't I just figure out a way to get it to absorb yellow as well? Wouldn't that hide the banana from you?

Sort of. The color black absorbs all light waves, so you could just paint the banana black. While this would probably make the banana a little bit harder to see, it wouldn't exactly make it invisible, would it? No, of course not. Technically speaking, if the banana is covered in black paint, you cannot *actually* see the banana. What you can see is a bunch of black paint in the shape of a banana, which is, strictly speaking, different from actually seeing the banana itself. We have succeeded in making the banana disappear, but we certainly have not made it invisible. If that banana were holding a rocket launcher and getting ready to fire it at you, the black paint wouldn't do very much to stop you from seeing it, would it? No. But why not? Because, even if the banana were painted so black that it absorbed every possible light ray, and even if, in addition to covering the banana with black paint, you somehow changed the shape of the banana so you didn't know it was a banana, you would still have the problem of not being able to see all of the stuff *behind the banana*. You might not know it was a banana, but you would certainly know that something was there because all of the light rays coming from your light would be stopping on this black blob in the middle of the table, which isn't really the same as being invisible.

What you want is for the light waves to leave the flashlight, travel either around or through the banana, touch all of the stuff behind the banana, bounce back again toward you, once again pass through or around the banana, and then hit you in the eye. If that were to happen, then all you would see is the table behind the banana. Of course there is a slight problem with this scenario: light cannot travel through a banana. At least not visible light (X-rays can pass through a banana, but our eyes don't see X-rays, so that's not helpful).

Thus, if we can't get the light to pass *through* the banana, we just have to do the next best thing: we have to make the light travel *around* the banana. Not only that—it has to travel *perfectly* around the banana, so that every ray of light that touches the banana is bent around it and then exits on the exact opposite side from where it first came in contact with the banana.

This concept might seem impossible, but it is actually behind all of the current technologies leading toward some kind of real invisibility. Some scientists have actually managed (albeit on a microscopic scale) to make light do exactly this (though, to my knowledge, not with a banana).

But how did they do it? How can anybody create something that bends light so perfectly around a physical object and makes it exit perfectly on the opposite side?

This is exactly the problem that scientists and engineers are trying to solve, and though they haven't quite fully succeeded yet, they have made a good bit of progress.

There are a few different ways that scientists have managed to bend light in this fashion. The simplest way is with a series of lenses. Lenses, like the ones in your glasses or a telescope or a magnifying glass, are precisely shaped pieces of glass that bend light in precise ways. A convex lens will bend light to make things appear larger, and a concave lens will bend light to make things appear smaller. Lenses can make light appear sharper or more blurry, and they can even make light simply stay the same, but appear as if it has moved slightly in one direction or another.

Recently, researchers at the University of Rochester figured out an exact configuration of lenses, both in front of and behind an object, so that, as long as you were looking at the lens, the object between the two lenses would not be visible at all, but the background would be perfectly visible.

Essentially, this is accomplished with a simple mathematical calculation wherein the lenses are all pointed in exact ways so as to move the light in the precise direction it needs to be so that anything in the "field of invisibility" cannot be seen. Their efforts with this method have been very successful; however, the method itself has some obvious drawbacks. First of all, the invisibility is only seen from a very limited viewing angle. If you move to the side and look perpendicularly between the lenses, you will see the object just fine. Also, and more obviously, you need a whole bunch of lenses placed in precise positions to make this method work at all. This means that the invisibility is quite limited to a very specific area between the lenses. It's a pretty cool concept, but not very useful in battle, though it does apparently have some interesting medical applications.

Another fairly simple means for creating this type of invisibility effect, one that is being researched by a number of military contractors in particular, is the use of cameras and screens for "moving" the light around an object. If I were to place a screen directly between your eyes and a banana, and then place a video camera on the other side of the banana, pointed in the exact direction you were looking, and then display the image from the camera on the screen . . . you see where I'm going with this. You would be looking at a screen with an image on it of what was behind the banana. If the screen were very thin, or difficult to see, you could very well be tricked into thinking that there was no banana there at all.

Furthermore, as this type of screen and camera technology continues to advance, we have the ability to create more adaptable and flexible screens and cameras. So one could even imagine, for example, placing screens and cameras all over a tank or other military vehicle, with each screen showing an image from a camera mounted directly on its opposite side. This would take a fair bit of engineering to get just right, but it is certainly a doable solution for invisibility. That said, this method also has its drawbacks. First of all, it requires a great deal of power. Second, as good as screen technology has gotten, it would still be extremely difficult to make something like this work in a way that was undetectable to the human eye.

This brings us to the third solution, and the one with the most promise for making true invisibility a reality: metamaterials. Metamaterials are man-made materials, usually constructed at the microscopic scale out of different types of exotic materials that, when properly made, will bend light around the outside of an object.

But still, even with this amazing technology, could anybody achieve anything like what we see with the Shadow Bomb in Fortnite? Well, we can certainly say with a great deal of confidence that nobody has invented that kind of technology just yet.

Index

NOTES

NOTES

....

....

....

....

....

....

....

....

....

....

....

....

....

NOTES

NOTES

....

....

....

....

....

....

....

....

....

....

....

....

....

....

NOTES

....

....

....

....

....

....

....

....

....

....

....

....

....

NOTES

....

....

....

....

....

....

....

....

....

....

....

....

....

....

NOTES